TESTIMONIALS

Berry Fowler
Founder of Sylvan Learning Centers.
Berry Fowler and Associates, Inc.
Fowler School of Business and Executive
Coaching

"I always knew Lorne had a book in him! His deep insights into AI, combined with his message of 'Careful Thinking,' offer a powerful and timely perspective. For anyone seeking a clear and thoughtful exploration of AI, this book is an absolute must-read!"

Michael Lewis
Chief Executive Officer
Advisyn
GTM & B2B Growth Advisory, Executive Coaching &
Organizational Strategy
https://advisyn.com

In *AI: Careful Thinking*, Lorne Ray makes AI as easy to understand as chatting with a good friend. Using his 'Lorne-speak,' he breaks down AI's complex concepts in a way that's both relatable and insightful. His focus on 'Careful Thinking' cuts through the hype, revealing AI's true potential. It's a must-read for anyone wanting to understand and apply AI thoughtfully.

AI - Careful Thinking
What AI IS and is NOT
By: Lorne C. Ray, MBC, CAIC

ISBN: 978-1-937925-51-2 Paperback
ISBN: 978-1-937925-52-9 eBook

Published by:

Bookjolt
An Imprint of Publishers Solution, LLC
4956 Long Beach Road Suite 14
Southport, NC 28461
www.bookjolt.com

ACKNOWLEDGEMENT

I want to mention a few people who have inspired me to write this book. I believe it is important to recognize those who are the "building blocks" that provide encouragement and hope in my life.

Pastor Allen Jackson in Murfreesboro, Tennessee, is what I refer to as a "man of common sense" who, although a man of God, embodies a spirit of kindness and concern for all. He is a "careful thinker."

Dr. John Lennox is another person I want to acknowledge. He is an Oxford mathematician and a tremendously careful thinker. He has taught me to see beyond the argument of science and faith and recognize the true meaning of ethical understanding.

I want to thank Bill and Nancy James (my book publishers), who believed in me and my message long before this book existed. They have used their decades of book publishing experience to help "fine-tune" the message, and I am so grateful to them.

Last, I want to thank my wife but not in a syrupy way. Elisa is tough. She is street-smart. She is a survivor. Never was there a more careful-thinking person than she is.

As you read this book, please know that these people are what I call "my building blocks of inspiration," and they all have made me a more careful-thinking person. Artificial Intelligence is clearly part of the equation in this book after all, it is in the title. But, the common denominator for the equation is certainly careful thinking.

TABLE OF CONTENTS

Get Familiar with McClear

Hi everyone, my name is McClear, and I am Lorne's chatbot. Chatbots exemplify AI's practical application in daily life, serving as an accessible interface for human-machine interaction while driving advancements in natural language processing and data-driven learning. In Lorne-speak, I help him understand things better, faster, and with more perspective. This is how it works: he asks me a question called a "prompt," and then I go out to the outermost parts of the Cloud and gather information that fits that prompt. I share with him what they have to say and I can do that in all kinds of useful ways. It could be a book, a marketing action plan, a business summary, PowerPoint slides, or more. Then, it is up to Lorne to decide how to best use it.

Lorne is pretty big on "careful thinking," as you know from the title of this book!

Oh, one last thing: you can name your chatbot whatever you want! Make sure, though, that you give them the credit they deserve; it is the ethical thing to do!

PROLOGUE

Question: I am an average person without a tech background. I have used email, text, social media and apps. I have played around with SIRI and Alexa and yes, I have used ChatGPT. AI, according to most experts, is the coming wave. Many say if you don't catch this wave, you will be left behind in the dust. So, why should I get on the wave, how do I get on it and how do I stay on the AI surfboard, safely? This book is for all of us who don't live in Silicon Valley, we drive a Honda, and do our own grocery shopping. If someone asked me the overall theme of the book my answer would be the following: Use "careful thinking" when it comes to AI. Period.

At this point in time, AI is like a slow-moving merry-go-round. It is moving and if you can time it right, you can jump on the merry-go-round with relative ease. Every month that goes by, AI is making that merry-go-round faster and faster. By the time you read this book, the merry-go-round will be moving significantly faster and faster and faster.

I realize that for many, AI seems very complex and hard to wrap your head around. The objective of this book is to help you make sense of it all. Granted this is a big challenge because of the promises and grandeur of AI. When you are done reading this book you are not going to be an expert at AI. You will, however, have your eyes opened up to the importance of being very careful and think about AI in a new light hopefully.

Remember,
"Our minds are the control towers of our thoughts."

Careful thinking is required with Artificial Intelligence!
To help make it easier to understand, I decided to break AI down into five parts and then throughout the rest of the book have some fun with where this is all going.

1. **Narrow AI:** Where AI specializes in one specific task. Examples include chess programs, image recognition software, or virtual assistants like Siri and the more advanced chatbots like ChatGPT4o.

2. **General AI (AGI or Artificial General Intelligence):** Where AI would be able to perform any intellectual task a human can. This doesn't exist yet, but it's a goal scientists are working towards.

3. **Super AI:** A hypothetical AI that would surpass human intelligence across all fields. This is still in the realm of science fiction.

4. **Machine Learning:** A subset of AI where systems improve their performance on a task through experience, without being explicitly programmed.

5. **Deep Learning:** A more complex form of machine learning using neural networks, often used for tasks like speech recognition or autonomous driving.

Narrow AI is the most common form today. While very capable in its specific domain, it lacks the flexibility and broad understanding that characterize human-like intelligence. The objective of this book is about finding value using AI. I am more encouraged by its potential and more cautious about its risks. By now I am certain that you get the message; Careful thinking is my prevailing message.

This book will reveal most, if not all, of the reasons I believe that those who get on the AI merry-go-round will thrive...but get on with careful thinking.

INTRODUCTION

About three years ago I started to hear more and more about Artificial Intelligence (AI). A friend of mine ("Dr. Scott" Dell, DBA.) began playing around with it and I thought he just had too much time on his hands. Dr. Scott is my "AI running-mate" and colleague. Like Butch Cassidy and The Sundance Kid we are both very different and come from radically different backgrounds. One thing though, that connects us like glue, is we both believe in the potential of AI and the possible risks it presents.

I would be remiss not to mention another colleague of mine, Liz Venendaal. She is the owner and CEO of FIA Coaching, an internationally recognized leader in the life and business coaching world. Collaborating with Liz in terms of the future impact that AI will have on coaching has been incredible. The unique aspect about Liz is that she thrives despite having cerebral palsy. She has a beautiful outlook on life and has turned her disability into a remarkable "ability" and inspires people beyond measure. She is incorporating AI into her coaching company and the possibilities for this are amazing.

Chances are, you have read about AI in the media, played around with Siri and Alexa, or Google Maps, and some of you have even used a chatbot. You may be both fascinated and a bit freaked out by it all. But like most humans, you are curious.

Curiosity is where I would like to begin, as it is the driving force behind me writing this book. I looked around the Internet for books on AI and they were all pretty techie or philosophical. None really got to the heart of the matter. How do I find my "aha moment" or what I call my Value Moment with AI? The tech driven books like "The Coming Wave" by Mustafa Suleyman (an excellent book by the way) target the Silicon Valley types. Others, like "Nexus" by Yuval Noah Harari, present some incredibly frightening aspects of AI.

My position on AI is pretty simple. A hundred years ago I would not have told you to avoid owning a car or buying a radio. I would have told you to apply **careful thinking** when using them. Throughout this book my overall theme is based on the fact that AI moves information much faster than we have ever experienced. Like the car, train and airplane moved us faster or the phone, radio and television got information to us faster, AI is doing the same. Be careful.

In addition to the excitement around AI, I figured there were millions of other people, hundreds of millions in fact, who share my sentiment of being careful. When I heard the saying that "AI levels the playing field forever," that got my attention. Then I heard a writer named Seth Godin comment that "AI is the biggest thing since electricity" then that got my attention even more.

If you are looking to understand AI more, and even find your entry point into the opportunities and doors that AI is unlocking, this book will hopefully help you make that step. My journey into AI started with what it is, what it can become, and how I fit into it all. I then looked at the risks and the ethics of AI.

Like any good writer I began doing a little local market research by talking to my friends and family. What I heard were words like; cool, weird, creepy, ethics, big brother is looking in, manipulation, fascinating, mind control, context vs. pretext, then creepy again and again.

Let's face it, the idea of artificial intelligence is creepy. How could a machine know what I like and need more than I do? None of us want robots to take over the world. We have seen way too many movies I guess. To many of us, AI is like; iRobot, The Matrix or The Terminator...inherently bad. Initially, I felt the same.

What I discovered and hope to share with you in this book, is that AI is more about efficiency than evil. Perspective and careful thinking should be the common denominator with AI being merely a tool. Fire has proven to be an amazing tool once we learn to use it effectively. Electricity, water, and wind are the same. It is about harnessing the use of something that makes it valuable to us as a society and individuals.

By harnessing something, we find its value and that "aha" moment that I call your Value Moment I certainly hope this book helps you find yours with AI.

Let me tell you a bit more about me. I was born to parents who had roots in farming and logging in Oregon. Both were non-educated for the most part. My mother got her GED in her 60's and my father went to college for a few weeks and then left to be a tree-faller and log truck driver. My mother is my hero and my father (who committed suicide at age 69) has become a bit of a mythical figure in my life. I choose to remember him as the dad who won the big logging show when I was a kid. He was my hero then and he taught me things early in life that still serve me well today. I have an older brother (who died of a heart attack at age 42) and a younger sister. We were a middle-class family through and through.

Back to my mother. Her best advice to me ever were the words, "Lorne, you can be anything you want to be in life if you want it bad enough." Add to this, time after time when I failed, she always said, "you're resilient, you will get back on your feet!" And I always have.

Like many Americans, my life unfolded by graduating from high school, getting married, raising a family, getting divorced, getting remarried...going from one career to the next, starting a business, and another...dealing with tragedy...fighting financial hardships, making mistakes and well...living life.

Then one day, I was 65! This is where my Value Moment begins. This book is about AI and not my faith (anyone who knows me, knows you probably don't want to get me started on my faith...that's another book), but AI impacts my faith a great deal.

My faith is fundamentally rooted in two words, value and serve. As you read this book and learn more about AI, you will learn to understand the importance of finding your own Value Moment.

Throughout this book you will see first-hand how I partner with AI to add value and serve you as the reader more effectively.

Enjoy this book in terms of simply gaining more perspective on AI. I will explain the details of AI, share some personal stories that helped me with my AI journey and hopefully help you see and experience AI as an asset in your future. Remember, use **careful thinking**.

CHAPTER ONE:

WHAT IS AI?

In this chapter you are going to learn why I refer to AI as "accelerated intelligence" versus "artificial intelligence." Like the jet engine accelerated air travel, AI is accelerating information.

To fully understand what AI is, you need to start with why it was created in the first place. In what I call "Lorne-speak" it was created to process information faster and hopefully more accurately. Dr. Scott and I often refer to AI as "accelerated intelligence." AI accelerates the rate in which we get information and then helps us put it in a format that is easy to share. Pretty cool.

It is like air flight in many ways. Orville and Wilbur Wright had a fascination with the idea of speeding up travel. If people could fly versus drive it would be faster. After all, whether you walk, drive or fly, when you get to where you are going, you are there. By flying you could get there faster. Flight simply accelerated travel.

As I mentioned earlier, throughout the book you will see "Lorne-speak" occasionally. It is my way of putting everything into language so we can all understand and relate more easily. In the process, you will be introduced to my chatbot AI buddy. His name is McClear. Together, we will have an ongoing dialogue to help make AI more understandable and useful. Hence, the name Mc "Clear."

In many ways, AI is the airplane (or rocket if we are honest) to processing information. The faster we are able to process information, the more information we can discover. Kind of cool. When you think about it, AI helps us get smarter faster. In a similar fashion, we go to school, college and then seek higher education degrees to help us know more and get further ahead in life, faster.

But, like the development of the airplane, where there was a great deal of trial and error, AI will go through a similar process for us as individuals as well. First of all, when it comes to flying, you cannot fly a plane, jet, or rocket (unless it's a drone), without getting in the cockpit and knowing how to use the instruments of flying. Unlike the early days of flying when Orville and Wilbur had no instructor and just took their chances, when it comes to AI, look at this book as your flight instructor.

Admittedly, AI is so new and so evergreen, it is going to be difficult for any instructor to know everything. In the early days of flight, a flight instructor only needed to know very rudimentary flight instructions because there was only a single engine on the airplane with very simple instruments to master. Learning AI is like a flight instructor having to know how to teach you to fly single engines, twin engines, jets, rockets, and drones all at one time!

Pump the brakes! My goal by the end of this book is that you learn the fundamentals of AI and help you implement it in a healthy perspective. In my imagination you might be curious, fascinated, and a bit confused about AI. It is the curiosity and fascination though, that drove the Wright brothers to flight. I hope I can help you convert your curiosity about AI into a powerful tool and add value to your life.

So let's get started and begin with what AI actually is, why it was created, and then hopefully by the end of this book, you are ready to book your AI flight. One final note, please keep your seat belts fastened and enjoy the fight!

AI was created for several key reasons:

1. **To automate and enhance complex tasks:** AI systems can process vast amounts of data and perform complex calculations much faster than we can. AI is valuable for tasks like data analysis, scientific research, and industrial automation. Compare the original airplane the Wright brothers flew to a modern fighter jet. Both fly, one just does a **lot more by using automation to perform faster.**

2. **AI can help solve difficult problems:** AI techniques like machine learning can uncover patterns and insights in data that we might miss, potentially leading to breakthroughs in fields like medicine, climate science, and economics. In a similar fashion, modern aviation has technology on board like radar and 3D imaging that enables the pilot to see more than the original pilots only saw with the naked eye. Oh, my buddy McClear will help us understand big words like "machine learning" as you read on.

3. **AI can enhance our capabilities:** AI chatbots (like McClear) and tools can help us be more productive and creative by handling routine tasks, providing information, and offering new perspectives. Features like "auto-pilot" enable modern pilots to sit back and relax while the aircraft flies itself. AI is kind of like learning to use the "auto-pilot" feature that helps us to see and hear things that we might miss along the way.

4. **To help us improve decision-making:** In fields ranging from business to government, AI systems can analyze complex scenarios and provide us with data-driven recommendations. In a similar fashion, modern technology enables a pilot to fly around storms and away from turbulent weather.

5. **To make money by creating new economic opportunities:** Development of AI technologies has spawned new industries and job markets. As aviation advanced and planes could carry more capacity, new industries were created like FedX and now SpaceX.

The ultimate goal of AI research is to create systems that can perform tasks that typically require human intelligence and time, potentially leading to transformative changes in many aspects of society and human life.

In "Lorne-speak" AI was created to automate, solve, augment, advance, explore, improve and create. I like these words, and in my imagination, you do as well. I will get to these in a couple paragraphs but first let's address ethics.

In the introduction to this book I talked about ethics and this certainly plays a central role in the use of AI. You simply cannot get too far into the subject of AI and not address the 800 lb. elephant in the room relative to the negatives of AI. When you talk about what AI is, you really need to address what AI should not be. AI is a learning machine. The machine learns based on what you tell the machine. Based on input into AI, it can occasionally generate information that seems plausible but is actually inaccurate or completely fabricated. It's like when a person remembers something incorrectly or makes up details; the AI is producing responses that it "thinks" might be correct based on data input, but in reality, is not based on truth. It potentially can and will give you bad information.

We have seen inaccuracies like this first hand with social media in the recent news. By their own admission some of the household names in social media have admitted to entering false data into their platform regarding Covid 19. Their reasoning was due to pressure from the government. Regardless, bad or incorrect information will result in misinformation.

16

In a similar manner, we will certainly see the same with AI. Some humans are good and some humans are bad. Some have different perspectives and experiences so their opinions and input into the machine will be reflective of their own beliefs. It is called bias. Here is what McClear has to say about bias: **AI systems can perpetuate and amplify existing societal biases if not carefully designed and monitored, potentially leading to unfair or discriminatory outcomes in areas like hiring, lending, and criminal justice.** Keeping things in context is very important when using AI. More about ethics later in the book.

Back to the seven words that I mentioned a couple paragraphs back. They are, automate, solve, augment, advance, explore, improve and create. These ideas are very exciting to me. If I could learn to use AI and it does all of these things for me, I am interested. In fact, I am jacked up! You should be as well. Whether you are a parent, a teacher, an entrepreneur, or a doctor, if AI does these seven things for you, that is exciting.

All of the other social media platforms to date cannot do these seven things. Current social media platforms help us share information, tell us what something is and enable us to explore deeper into things. But, they don't automate, solve (or at least present solutions for consideration), to improve, or create.

Let's stop for a minute and examine the word *create*. If you ask Google to create a course for a particular subject it takes you to sponsored ads first, then it gives you YouTube videos on "how to" create a course on that subject, then it gives you courses you can take on that subject, then it takes you to the "People Ask" section and then you get results about a definition of that course. That's it. With AI (using a chatbot like ChatGPT4o or my chatbot McClear) you literally get a course outline and then it asks you if you would like it to expand on each section. When you do, you have a course! It can literally create an entire course. *(By the way, let me tell you a little secret. If you use a chatbot like ChatGPT4o or Claude 3.5, you can rename it any name you like. I like McClear. A chatbot is like a pet, just because it is a dog, you don't have to call it "dog." You can call it Lassie or Rin Tin Tin!)*

Next let's use the word **improve**. If I ask Google or people on a social media platform to review and improve a particular course I have, Google will send you to ads on people or companies who charge you money to improve your course, then to YouTube for the same, and then social media platforms like Facebook and LinkedIn taking you to people or companies who charge you money to perform this task. A chatbot gives you improvement suggestions for free. And, in my experience, the suggestions for improvement far exceed the ones I get from search engines or on social media. I think sometimes, *McClear thinks he is underpaid and over-worked. Too bad McClear, good for me! But, be sure McClear, I will always give you the credit you deserve for your assistance. Thank you.*

AI's ability in the other areas like problem solving, automation, augmenting, advancement, and exploration are off the charts compared to any search engine or social media platform. You see, in Lorne-speak, AI is like having the smartest person in the world sitting by your side and available to help you 24-7. I said smartest, I didn't say the kindest, most loving, caring or compassionate person. (I have someone else for this and that's another book). In Lorne-speak, think of AI kind of like the Wizard of Oz. The hope that Dorothy had all along was that the Wizard provided her with good advice. But in the story, he was a conman. Big-tech to a certain extent can be the Wizard of Oz in your life, if you let it, or it can be the "Wizard of Good" if you have the right perspective and use **careful thinking**.

The more I researched for this book, the more I came to understand what AI is and what AI should not be in our lives. Throughout the book I use many different illustrations to drive home the point that, if used properly, AI can be wonderful. Having the smartest person by your side who gives you good information and can formulate it into any format you ask is valuable.

If I were asked to use words that describe AI to me, they would be; smart, fast, thoughtful, thorough, efficient, time saving, powerful, helpful, clarifying, organized, and complete. Think of these in this light, no matter what you do, if these words were all incorporated into everything you do personally and professionally, would you be better off? I think you would. Value Moment! But, always remember these two even more important words; **careful thinking**.

CHAPTER TWO

WHY IS AI VALUABLE?

> **In this chapter you will see examples of how AI can help us with personal issues and business issues with clear and succinct results.**

Something that is useful or has a practical purpose is often considered to be valuable. This could be something that helps us solve problems, make our lives easier, or improves our health or well-being.

With *useful* as the context for value regarding AI, the question then becomes, how do I use AI to be useful in my personal or business life? AI has emerged as a transformative technology with many valuable applications to aspects of our lives both personally and in business. Essentially, it has the ability to process a lot of data, learn from experiences by recognizing patterns that we often overlook, and deliver these results in a context that we can then decide if it is useful.

In Lorne-speak, it sees what we don't see and hears what we don't hear and processes results lightning fast. AI then assimilates this information in any format we ask for our use. So, I asked McClear to give me an example of something relative to personal and business use to demonstrate its effectiveness.

Personal Use: My wife likes arts and crafts, wants to be healthier and likes live music. I ask McClear to write me a poem that she might enjoy based on these few criteria. Here is what I got back from McClear:

My Elisa

With hands that craft and eyes that see, You shape the world so artfully. In every stitch and brush's stroke, Your passion for creation spoke.

A melody of wellness rings, As healthier habits spread their wings. Your strength grows with each passing day, In mind and body, come what may.

When music fills our quiet home, Your soul, it dances, free to roam. The notes that float on evening air, Remind me why I'm glad you're there.

In twilight's hush, just you and I, No need for words as time slips by. Your presence is a soothing balm, These quiet evenings, full of calm.

Elisa, love, in all you do, You bring such beauty, tried and true. In art, in health, in music's sway, You make me fall more in love each day.

Here is a spoiler alert. I write poetry and don't need a McClear to do this for me. But, for a guy who needs to step it up in his relationship, this is a pretty valuable tool for warming up his lady's heart. Even if he fesses up that he used AI to write it, he had to tell AI the things important to her and that is showing that he cares. This input is called **"prompt engineering"** by the way. More about prompt engineering later in the book!

Business Use: Let's say I am a life coach and I have a client who is struggling with time management. I asked McClear the following prompt: "My client Joe is doing well with everything in his life except for time management. Can you write me a program that will help Joe with time management?" McClear decided to show off a bit...here is what he came up with:

Sure! I can help outline a time management program that Joe can use to get organized and improve his time management skills. Here's a structured plan Joe can follow:

Joe's Time Management Program

1. Goal Setting and Prioritization

a. *Define Goals*

- Short-Term Goals: List what needs to be achieved in the next week or month.
- Long-Term Goals: Identify objectives for the next six months to a year.

b. *Prioritize Tasks*

- Use the Eisenhower Matrix to categorize tasks into:
- Urgent and Important
- Important but Not Urgent
- Urgent but Not Important
- Not Urgent and Not Important

2. Planning and Scheduling

a. *Weekly Planning*

- Weekly Review: Every Sunday, review goals and tasks for the upcoming week.
- Create a Weekly Plan: Allocate specific times for tasks based on priority.

b. *Daily Planning*

- Daily To-Do List: Start each day by listing tasks to be accomplished.
- Time Blocking: Assign specific blocks of time for each task to avoid multitasking.

c. *Use a Planner or Digital Calendar*

- Choose a Tool: Use tools like Google Calendar, Outlook, or a physical planner.
- Set Reminders: Schedule reminders for important tasks and deadlines.

3. Time Management Techniques
a. *Pomodoro Technique*
- *Work in Intervals: Work for 25 minutes, then take a five-minute break. After four intervals, take a longer break (15-30 minutes).*

b. *Time Tracking*
- *Track Time Spent: Use apps like Toggl or Clockify to track time spent on tasks to identify time drains.*

c. *Task Batching*
- *Group Similar Tasks: Batch similar tasks together to reduce switching costs and increase efficiency.*

4. Overcoming Procrastination
a. *Break Tasks into Smaller Steps*
- *Subdivide Tasks: Break large tasks into smaller, more manageable steps to reduce being overwhelmed.*

b. *Set Deadlines*
- *Create Mini-Deadlines: Set deadlines for each step to maintain momentum.*

c. *Reward System*
- *Incentivize Completion: Reward yourself after completing significant tasks or achieving milestones.*

5. Review and Adjust
a. *Weekly Review*
- *Evaluate Progress: Reflect on what was accomplished during the week and what needs adjustment.*
- *Adjust Plans: Modify goals, priorities, and strategies as necessary based on your review.*

b. *Monthly Check-In*
- *Assess Achievements: Review long-term goals and progress towards them.*

- *Plan Adjustments: Make adjustments to strategies and goals based on progress and any changes in circumstances.*

6. Tools and Resources
a. <u>Digital Tools</u>
- *Todoist: For task management and tracking.*
- *Toggl/Clockify: For time tracking.*
- *Evernote/Notion: For note-taking and organizing information.*

b. <u>Books and Courses</u>
- *Books: "Getting Things Done" by David Allen, "Atomic Habits" by James Clear.*
- *Courses: Online courses on platforms like Coursera or Udemy on time management.*

Implementation Steps for Joe

1. **<u>Set Up Goals:</u>** *Start by defining both short-term and long-term goals.*
2. **<u>Choose Planning Tools</u>:** *Select a planner or digital calendar.*
3. **<u>Adopt Time Management Techniques</u>:** *Implement the Pomodoro Technique and start tracking time.*
4. **<u>Create a Weekly and Daily Plan:</u>** *Establish a planning routine.*
5. **<u>Review and Adjust Regularly:</u>** *Perform weekly reviews and adjust strategies as needed.*

TM

Encourage Joe to start with one or two techniques and tools, gradually incorporating more as he becomes comfortable. This approach will help him avoid feeling overwhelmed and increase the chances of successful time management.

Not bad right! Just like that, I have a well written, organized and clear roadmap for Joe to implement regarding his time management challenges.

These are just two brief examples of how AI is so valuable. Both are thorough and organized. That makes the results useful. In both cases, personally with the poem and professionally with time management, AI delivered really clear results.

In many ways, AI does for us what we want to do but simply don't have the time or the complete scope of knowledge to get it done for ourselves. Clear and concise are two words that I associate with AI in terms of the value it provides. Still, be careful and think through whatever AI presents to you. It can make mistakes.

If you are like me, you can sometimes easily go off the rails with what you are trying to say. My intentions are good but my execution can sometimes get a bit fuzzy. Here is another example, I am going to ask McClear: How can I write in a more clear and succinct manner? Here are the results:

McClear was happy to oblige: To write more clearly and succinctly, a writer can:

1. *Use simple language: Avoid jargon and complex words when simpler ones will do.*

2. *Be specific: Use concrete details instead of vague generalizations.*

3. *Cut unnecessary words: Remove filler words, redundancies, and overly wordy phrases.*

4. *Use active voice: It's often more direct and engaging than passive voice.*

5. *Organize thoughts logically: Present ideas in a coherent sequence.*

6. *Break up long sentences: Shorter sentences are usually easier to understand.*

7. _Use bullet points or numbered lists:_ They help organize information clearly.
8. _Revise and edit:_ Always review and refine your writing.
9. _Read aloud:_ This helps identify awkward phrasing or unnecessary complexity.
10. _Get feedback:_ Ask others to review your writing for clarity.

Thank you, buddy. I needed that! Value, value, value. In real time, using AI technology, McClear served as my instructor helping me to write more effectively. Let's address 'number 4' above. Use an active voice. What does this mean? Why not ask for examples of an active voice when writing. Here is what McClear had to say about this:

Certainly. I'll provide examples of active voice and explain why it's often preferred in writing.

1. "The chef prepared the meal."
2. "Students submitted their assignments."
3. "The committee approved the proposal."
4. "Sarah wrote an award-winning novel."
5. "The company launched a new product."

Active voice has several advantages:

1. Clarity: It clearly identifies who or what is performing the action.
2. Conciseness: Active sentences are often shorter and more direct.
3. Energy: It creates a sense of action and immediacy in writing.
4. Responsibility: It doesn't hide the subject, making it clear who's responsible for the action.
5. Engagement: Readers often find active voices more interesting and easier to follow.

In contrast, passive voice reverses this structure:

"The mouse was chased by the cat."

While passive voice has its uses, it can make writing feel indirect or evasive. For example:

Active: "The manager made a mistake." Passive: "A mistake was made."

The passive version obscures who made the mistake. End.

Thanks McClear! Often when we are writing using the passive voice, our words wind up being vague. Conversely, the active voice simply provides more clarity. Being clear is critical if you want the maximum understanding of what you are saying to come across to your audience. To learn this though, I didn't have to take a refresher course at the local community college, I simply had to ask my AI buddy McClear. That is value.

CHAPTER THREE

HOW TO USE AI?

In this chapter you will see why it is so important to have good instruction when it comes to using AI. I will use the illustration of a pilot having a flight instructor and learning to use the instruments on the plane.

I kind of let the cat out of the bag on this one in the last chapter. But, using AI goes much deeper. In Chapter One I drew the analogy of AI to air flight. It is interesting to note that all pilots start out in a glider or single engine airplane. In fact, if you are reading this book flying across the ocean on a huge airliner, the captain who is piloting the airliner once started out in a little tiny single engine plane! He/she got a pilot's license, then learned to use the instruments on the plane, then got their Instrument Flight Rating so they could fly at night or in the clouds and then more and more ratings until they eventually could fly the airliner you are a passenger on.

In this chapter when we talk about how to use AI, the goal is to give you the basics of AI but you are probably not ready to get in the cockpit with an instructor.

Here is the short list of how to use it:

1. **Choose an AI tool or platform** (e.g. ChatGPT, Claude, Gemini, etc.) based on your needs. Of course, these are chatbots. Download one of them and simply make a request. It's kind of fun actually. As you know by now, mine is named McClear.

2. **Familiarize yourself with the AI's capabilities and limitations.** Understand the fact that AI does make

mistakes. Ouch!! I know that kind of stings McClear.

3. **Formulate clear, specific prompts or questions.** The more specific you are with your input, the more accurate (hopefully) AI will be in the response you get back.

4. **Provide context and necessary information.**

5. **Evaluate and verify the AI's responses.** Time to think carefully!

6. **Iterate and refine your prompts as needed.** If you don't like the results you received, ask it a different way. Even if you do like the results, ask for elaboration if you desire.

7. **Use the AI's output as a starting point or aid**, not as definitive answers.

8. **Be aware of potential biases or inaccuracies** in AI-generated content.

Ok, that is a pretty cool list of how to use AI. I asked McClear to elaborate a bit more. (Look at your AI chatbot as your classroom instructor or tutor who is teaching you all of the basics of flying.

In the next section of this book we are going to get into the nuts and bolts of AI. Let's see if we can make some sense of it all. McClear will help us navigate through it all.

Here is a detailed guide on how to use AI most efficiently, covering the selection of tools, data management, implementation strategies, and continuous optimization provided by McClear:

1. Defining Objectives
1.1 Identify the Problem
Clearly articulate the problem you want to address or the opportunity you want to capitalize on with AI. This might involve automating routine tasks, improving decision-making, or uncovering insights from large datasets.

Ok, time for a little Lorne-speak. Let's face it, most of us are not going to be dealing with "large datasets." But in time, you might be surprised! Look at large datasets like having a bunch of really smart people in a room with you. They all know a great deal about particular subjects. But the problem is, they are not that easy to talk to because they are full of knowledge and information but they are not the world's best communicators! It's kind of like this; you ask them a question and they hand you a stack of papers with thousands of spreadsheets and tell you, here is your answer. Before AI, you would say, "ok...but where is it in all of this mumbo jumbo? I thought I clearly "articulated" what I was looking for?

Well, look at AI technology as another person who shows up in the room who knows all of these really smart guys and is able to speak propeller-head language with them and can easily communicate with you. My smart guy is McClear. He literally becomes my information broker or agent and helps me negotiate my way through all the papers and spreadsheets (the large datasets) and is able to get me answers and...in addition to the answer hidden in these large datasets, is able to ask me even more questions relative to what is inside all of this information! Kind of like a master gardener who knows how to get the most out of the soil.

So in Lorne-speak, look at AI as your "McClear" who is able to be your propeller-head interpreter and help you "clear up" all the *gobbledygook* of information hidden out there in the cloud and make it most useful to you.

1.2 Set Specific Goals

Define measurable objectives for what you want AI to achieve. Goals might include reducing operational costs, increasing sales, or enhancing customer satisfaction. These objectives will guide your AI implementation strategy.

Ok, in Lorne-speak it is time to call in McClear! You lost me at *"measurable costs!"* Look at it this way as if you were talking to a 10-year-old: Measurable costs are expenses you can easily count or put a price tag on. Here are some examples that McClear provided to me:

1. *The number of dollars you spend on a toy*
2. *How many gallons of gas your parents put in the car*
3. *The price of a movie ticket*
4. *How much a bag of groceries costs*
5. *The amount you pay for a haircut*

These are all measurable because you can see the exact number or price. It's like counting how many cookies are in a jar - you can only give it a specific number.

On the other hand, some costs are harder to measure, like how much fun you have playing with friends or how much you learn in school. These are important too, but they're not as easy to put a number on.

What AI does, once you have accurate numbers, is to prepare a much more accurate response in correlation to these costs. If you don't provide the right input, AI cannot give you the most accurate output.

Thanks McClear...that helps.

2. Choose the Right AI Tools
2.1 Assess Your Needs

Determine the type of AI technology that best fits your objectives. Common types include:

- *Machine Learning (ML): For predictive analytics and pattern recognition.*
- *Natural Language Processing (NLP): For text analysis and language understanding.*

- *Computer Vision: For image and video analysis.*
- *Robotic Process Automation (RPA): For automating repetitive tasks.*

Machine learning and natural language processing go hand in hand with AI from the beginning. Using chatbots is where you will begin with combining both of these. Computer vision and robotic process automation will come later, but in time they will be very important.

Lorne-speak time...McClear...are you there? *Machine learning* and *natural language processing*...are you kidding me? It's not that complicated. Imagine you're teaching a robot to play a game. Instead of giving it a list of rules, you show it many examples of how the game is played. The robot watches and learns from these examples, getting better each time. That's like machine learning!

Here are some everyday examples:

1. **Smart speakers:** They learn to understand your voice better the more you talk to them.

2. **Video game enemies:** They can learn from how you play and become more challenging.

3. **Movie recommendations:** Streaming services learn what kinds of movies you like by looking at what you've watched before.

4. **Spam filters in email:** They learn to spot junk mail by looking at lots of examples of good and bad emails.

5. **Face recognition on phones:** Your phone learns to recognize your face by looking at it many times from different angles.

In each case, the computer isn't just following a set of rules. It's learning from data and experiences, just like you do when you practice a new skill. Ok, McClear...that's kind of creepy. Not really Mr. Lorne. (can I call you this?) ...of course. Chances are you have Netflix or YouTube on your television. Both Netflix and YouTube have been using machines (computers, basically) to deliver the programs you selected. The more you watch, the more they know the kind of programs you enjoy. In time they are able to make suggestions on other shows you might like based on your own interests that their machines can keep track of. It is not really creepy at all, it kind of makes sense. AI just took this capability that these computers have and took it to the next level.

Ok, McClear...I am good to go with machine learning but help me get past this whole "natural language processing" idea. Easy Mr. Lorne...hey McClear...my friends call me Lorne. *Ok, Lorne; Natural language processing is teaching computers to understand and use human language, like English or Spanish, the way people do. Texting your friend: Imagine if your phone could understand exactly what you mean when you text, even if you use slang or make spelling mistakes. That's what natural language processing (NLP) tries to do. So here are some examples that might help you understand it even more:*

1. **Talking to a smart speaker:** When you ask Alexa or Siri a question, they use natural language processing (NLP) to figure out what you're asking and how to answer you.

2. **Autocorrect and suggestions:** When your device suggests words as you type or corrects your spelling, that's NLP in action.

3. **Language translation apps:** These apps use NLP to change words from one language to another, trying to keep the meaning the same.

4. **Homework helper:** Imagine a computer program that could read your homework questions and help explain the answers, just like a tutor. That's another way NLP can be used.

5. **Sorting emails**: NLP helps email programs figure out which messages are important and which ones might be junk mail.

The goal of NLP is to make computers understand language almost as well as you do. It's like teaching a robot to be really good at reading, writing, listening, and talking.

Ok McClear...kind of cool but still...that is pretty amazing. *Lorne, hang in there. It is artificial intelligence. NLP is driven by the word "processing." Computers simply process information faster than we can as humans. It is still based on human intelligence, AI just artificially processes our own information faster. Without human intelligence, the machines would simply sit there and rust away.*

2.2 Evaluate Platforms

Ok McClear...Let's talk about the rest of this AI stuff for a bit. *I'm ready Lorne. Ok McClear, what are some of the platforms?*

Select AI tools or platforms based on factors such as scalability, ease of integration, and support. Popular platforms include Claude 3.5, ChatGPT4o, Google's Gemini, Microsoft Azure AI, IBM Watson, and open-source libraries like TensorFlow and PyTorch.

Chances are, by the time this book is published, one or more of these platforms will have a new version or a name change. No worries, all you have to do is ask one of the other platforms what happened, they will love to share!

3. Data Management
3.1 Data Collection
Hey McClear, I am going to need some help on this data management stuff? *No problem Lorne:*

Gather relevant data that will be used to train your AI models. Ensure that the data is comprehensive, high-quality, and representative of the problem domain. Data is often a scary word for many of us. In simple terms, data is simply facts, numbers, names, figures or even the description of things. Generally data is organized in the form of charts, tables, or graphs.

3.2 Data Preparation

Clean and preprocess the data to remove errors, inconsistencies, and irrelevant information. This step is crucial for ensuring the accuracy and reliability of your AI models.

Pump the brakes McClear. I have to clean the data? Cleaning and processing data are like they do when you get off a flight. If you are one of the last to leave the plane, you see people come on board who start cleaning the plane and making sure there are no stowaways left on the plane! You want your data to be as clean as possible with no stowaways.

So for example, if you have a client list, make sure their contact information is accurate. Phone, physical address, email, web address, birthdays etc. Get the picture?

3.3 Data Privacy and Security

Implement robust measures to protect sensitive data and comply with relevant regulations, such as GDPR or CCPA. Secure data storage and access controls are essential for maintaining data integrity and confidentiality.

GDPR or General Data Protection Regulation, developed in the EU, safeguards the collection and protection of personal information. Google it!

CCPA is the state of California's version of GDPR.

I don't need McClear for this section. It is good to know that regulations are being put in place and refined as AI technology gets more sophisticated and powerful. It is good to know that companies and governments around the world are recognizing that sharing information and having safeguards in place make sense.

In the aviation world we all by now have heard of the "black boxes" that airplanes have on board designed to withstand plane crashes. These black boxes protect that important flight data to help aviation experts understand the cause of the accident. What is really powerful about the black box data is that airlines and governments have agreed to share this information with each other in an effort to make air travel safer for all. AI is following this model.

4. Implementing AI
4.1 Model Training

Train your AI models using the prepared data. This involves selecting appropriate algorithms, tuning hyperparameters, and iterating to improve model performance. Use techniques such as cross-validation to assess model accuracy. Don't worry about the technical processes at this point. AI can help you accomplish all that you require.

Ok McClear, "Tuning hyperparameters!" *Seriously? Lorne, when you board an airliner, there are all kinds of "thing-a-ma-jigs" that make that baby fly over 500 miles per hour. Algorithms and hyperparameters are "thing-a-ma-jigs" that make up AI. When you board an airplane do you go up front and expect the pilot to explain every part on the entire plane? Trust me Lorne, right now it's nothing you need to completely understand. If you did, you probably wouldn't be reading this book!* Alrighty then…let's move on McClear!

4.2 Integrating
Integrate the AI solution into your existing systems and workflows. Ensure compatibility with current technologies and processes to facilitate a smooth transition.

McClear...help! *Lorne, if you live in a cold climate you probably need a warm jacket. You need to make sure the jacket you select works best with the majority of your wardrobe. Look at AI as that warm jacket. You want to make sure the jacket is compatible with your shoes, belts, and slacks.*

Ok McClear...makes sense. *Hey Lorne, for the average user of AI, their current smartphone or laptop/iPad will be just fine. No worries.*

4.3 Testing
Conduct thorough testing to evaluate the AI system's performance in real-world scenarios. Identify any issues or limitations and make necessary adjustments.

McClear, you lost me here. I have to test this crap? *Lorne, when Boeing introduces a new aircraft into its fleet of planes, both Boeing and the airline do extensive testing to make sure the aircraft is safe and a good fit for the passenger load, time, and distance. Likewise, you will want to do some testing and evaluation of AI for your usage as well. All you are going to do is ask your AI chatbot (like you do me) to provide you with some preliminary answers. See if what they share with you makes sense. That's all. Don't get too worked up at this point.*

5. Monitoring and Evaluation
5.1 Performance Tracking
Regularly monitor the performance of your AI system against defined goals. Use metrics such as accuracy, precision, recall, and ROI to assess effectiveness. In other words, know where you are starting so you can measure the results of AI to determine its true value.

Ok McClear...more work for me? *Lorne, just pay attention to the information you receive. AI is designed to give you things for consideration essentially. Performance tracking is like watching the cake bake as it is cooking to make sure it is rising and then when you stick the toothpick in the center to see if the cake is done, you are just performance tracking. Not that difficult.*

5.2 Feedback Loop

Establish a feedback mechanism to capture user experiences and outcomes. Use this feedback to make iterative improvements to the AI system. Ok McClear, I think I am catching on. In fact, this is really cool. Ask a chatbot to create a survey for your customers based on what you used AI for and learn in real time from them their user experience! It's not rocket science after all. *No, Lorne...I told you...it all makes sense. AI is just there to help you see more and process things faster.*

5.3 Continuous Improvement

AI systems require ongoing refinement and updates. Continuously feed new data into the model and adjust algorithms based on changing requirements and emerging trends. When you get feedback from user experience surveys, simply ask AI to provide a detailed response to the feedback you received.

In Lorne-speak, feedback is really education. You are simply educating yourself on this feedback to serve your clients better. Better service means happier customers. That's a good thing! We are getting there McClear!

6. Ethical Considerations
6.1 Bias and Fairness

Be mindful of potential biases in your AI models. Ensure diverse and representative data to avoid perpetuating existing inequalities or biases. Use common sense.

Hey McClear, I have heard about this whole bias and fairness issue quite a bit. What's up with that? *It's pretty simple Lorne, if most of the big data in the cloud was created by a certain demographic, it is pretty logical that the output is going to be impacted by their input. Just be mindful. Over and over in your book you mention "careful thinking." I agree with you on this, especially when it comes to bias and fairness.*

6.2 Transparency and Accountability

Maintain transparency about how AI decisions are made and ensure accountability for the outcomes produced by AI systems. This builds trust and ensures ethical use of AI.

Hey McClear, I try to credit AI in all of my use of it whether I am working with a client or writing this book. Chances are, I missed a few...my chatbot buddies are pretty forgiving. I even give them a lot of credit on things they don't help me with. We are getting to be buds! Like you McClear!

7. Conclusion

Using AI efficiently involves a strategic approach from defining clear objectives to continuous monitoring and improvement. By selecting the right tools, managing data effectively, and considering ethical implications, organizations can maximize the benefits of AI and achieve your desired outcomes. Embracing a proactive and iterative mindset will help in navigating the complexities of AI and ensuring long-term success.

Ok, in Lorne-speak, that is a mouthful. But, by the time you finish this book it won't be so overwhelming. AI is something you will warm up to pretty quickly. It is kind of like riding a bicycle. At first you will feel a bit wobbly. But, in time, your confidence will grow and you will be rolling down the AI highway.

CHAPTER FOUR

HOW NOT TO USE AI

In this chapter you will see the dangers of using AI incorrectly. I will also talk about the hucksters who are already telling you how you can get rich overnight with AI. Be careful.

There are many potential pitfalls using AI. I can only imagine that in the early days of flight, there are many people who wish they knew ahead of time the pitfalls of flying. The idea of "crash and burn" was very unforgiving in flight. You want to avoid the same with AI.

Below are the top seven things I would suggest you pay attention to and never do:

1. **Never rely on AI for critical decisions** ignoring human oversight. Trust your gut.

2. **Never spread misinformation or generate fake content.** A very bad idea.

3. **AI can be biased and inaccurate.** Get the point? Be careful.

4. **Do not rely on AI for human empathy or understanding.** It's a machine!

5. **If you can do something yourself, don't rely on AI.** Common sense, right?

6. **Never let AI replace your personal creativity or original thought.** Pretty sure that based upon the data available at the time, if AI was in existence when the Wright Brothers decided to fly, AI would have told them it was impossible!

7. **Do not use AI for illegal activities** or circumvent ethical guidelines. Duh!

In Lorne-speak, do the right thing when it comes to using AI. For example, if you have presented write up papers for years, chances are that people will recognize your work. Then suddenly one day, you begin presenting these beautiful papers and you act as if some writing Genie suddenly appeared and you are Ernest Hemmingway. Don't do it.

With AI, the do's definitely outweigh the don'ts, but the don'ts can be pretty ugly. And yes, there is no doubt there will be plenty of ugly muddying up the water. Even when writing a book about AI, it is tempting to have AI do the work and for me and I take the credit. For starters, publishers already have standards in place to detect the authenticity of someone's work. Sites like Grammarly (a site used to edit writing), have tools in place to even check for plagiarism. I have collaborated with AI in certain circumstances with this book but I give them (the chatbots like McClear) credit. Not giving credit would be like leaving New York City on a jet and arriving in Los Angeles six hours later and trying to convince someone I ran all the way. Dumb.

I remember a funny story that relates to this whole idea of not using AI correctly. About 20 years ago, I stopped by this glider port where gliders were taking off every few minutes. It was apparently a really good day for gliding because of the thermals in the sky. On the wall of the little building where the glider pilots would go in and sign up to fly there was a promotional sign that read, "learn to glide today!" The more I watched these amazing little aircraft without engines fly, the more fascinated I became. Finally I went inside. I went to the counter and said, "I want to sign up and fly today!"

Well...here is what happened next. I am 6'5' and weigh about 240 lbs. today. Back then I weighed close to 295 lbs. The first thing the guy at the counter asked me is how much I weighed. I lied and somehow, he knew it. 250 lbs. was the max. Rather than him telling me he knew I was lying, he simply said, "yeah, dude...these little planes do pretty crazy things when they are overweight...like dude...they don't make it back to the runway!" That's all he needed to say as I crumpled up my sign-up form and went back outside to be an observer.

As much as using AI correctly can be useful, using it the wrong way can be equally destructive. Had I chosen to lie about my weight, I might not be writing this book. Do the right thing with AI and your experience will be positive.

One last thing, there are going to be all kinds of hucksters telling you how to make millions using AI. Here is what my buddy McClear has to say about hucksters and AI:

There is indeed a concerning trend of "hucksters" or opportunistic marketers promoting get-rich-quick schemes centered around AI. Here are some key points to consider about the dangers of these claims:

1. *Unrealistic promises: Many of these promoters make exaggerated claims about effortlessly making millions using AI, which are often highly unrealistic.*

2. *Oversimplification: They tend to grossly oversimplify the complexities of AI technology and its applications in business and finance.*

3. *Lack of expertise: Many of these self-proclaimed "AI gurus" may not have genuine expertise in artificial intelligence or machine learning.*

4. *Expensive courses or programs: They often sell overpriced courses, seminars, or "exclusive" programs that promise to reveal secrets of AI wealth generation.*

5. *Predatory tactics: These marketers may use high-pressure sales tactics or prey on people's fears of being left behind in the AI revolution.*

6. **Neglect of risks:** *They typically downplay or ignore the very real risks and challenges associated with AI implementations in business.*

7. **False sense of ease:** *By promoting AI as a magic solution, they create a false impression that success with AI is easy and guaranteed.*

8. **Potential financial harm:** *People who buy into these schemes may invest significant money and time, often with little to no return on investment.*

9. **Misinformation spread:** *These hucksters contribute to the spread of misinformation about AI capabilities and applications.*

10. **Distraction from legitimate opportunities:** *The noise created by these get-rich-quick schemes can drown out information about genuine, albeit more modest, opportunities in AI.*

It's important for people to approach claims about making millions with AI skeptically, do thorough research, and consult reputable sources and experts when considering AI-related business ventures or investments.

Thanks McClear, I couldn't have said it better myself. Listen to McClear, these hucksters are one of the main reasons I wrote this book. Know AI, appreciate AI, use AI, and most of all, be honest and participate with integrity and you will have a wonderful experience.

CHAPTER FIVE

PRACTICING AI

In this chapter I refer to the concept of being an "outlier" as brilliantly defined in Malcolm Gladwell's book, "Outliers." Those who stand out and do exceptional things do one thing more than everyone else, they practice!

If you are like most people, the words "artificial intelligence," today referred to simply as "AI," both fascinates and scares you. We have all heard the talk about the machine taking over the world. Our first foray into this idea was the television and to a certain extent, the radio.

Eighty-five years ago, a broadcast of Orson Welles's *War of the Worlds* supposedly caused mass hysteria in America, as listeners thought Martians were invading New Jersey. There are varying accounts of the controversial incident, and it remains a topic of fascination, even today. Back when Welles's fictional Martians attacked, broadcast radio was considered a state-of-the-art technology.

Still, the Welles incident struck fear in many. How do we know if it is real or fake when we turn on our radios? In many ways, AI struggles a bit from what I call the "War Of The Worlds Affect." Others call it the "fake news effect!" We have come a long way since the days of early radio.

Still, since the first transatlantic radio signal was transmitted in 1901, radio has greatly innovated the way we communicate. This opened the door to new technologies that can be really good and really bad at the same time. It is really up to us as individuals to learn how to use and interpret technology correctly.

Before radio and television, people actually interacted with other human beings. With radio we had to listen and with television we had to watch. Listening and watching in reality are fundamental to the human experience. In a sense you could say that television and radio actually helped mankind to see and hear better. To a point though.

With the radio, people had to sit around the little box and quietly listen if they wanted to hear what was being said. With television, you had to literally stare at the box in order to see what was on the screen. So, even though many people today blame the downfall of Western society on television and radio, it has had its benefits to society as a whole. We learned a lot faster with both technologies even though they did put a damper on our face to face communications.

This book in many ways is about having a "change of posture" towards AI and all technology in general. Many writers and speakers talk about that moment in life when we have our "aha moment" where everything comes together and we do something spectacular. This is not what I experienced particularly when it comes to AI. I call it my Value Moment. This is key to my message in this book.

When I tell you that the word "value" might be the most important word in all languages, I certainly hope I have your attention and you want to read on. For starters, this book is about AI and its impact on adding value to your life, your job and your family. However, if we don't get a clear understanding of the word value, you could miss the message.

Every advancement in modern history, from the industrial revolution to the tech revolution, was driven by the value it provided to the end users. The value is generally relative to saving time and money. Let's face it, we live in a global marketplace and time and money drive pretty much everything.

One closing thought before we move on. You will most likely find a great deal of repetition in this book. I have done this on purpose so that you begin to get familiar with the more technical aspect of AI in various different lights. Repetition is important to learning for several key reasons. Here are the eight reasons that McClear provided for us:

1. Memory consolidation:
- Repeated exposure helps move information from short-term to long-term memory
- Strengthens neural pathways associated with the learned material

2. Skill development:
- Practice leads to improved performance and muscle memory
- Allows for refinement and correction of errors over time

3. Deeper understanding:
- Multiple encounters with information allow for exploration from different angles
- Helps identify nuances and complexities not apparent on first exposure

4. Confidence building:
- Familiarity with material increases comfort and reduces anxiety
- Boosts self-efficacy in applying knowledge or skills

5. Overcoming the forgetting curve:
- Counters natural tendency to forget information over time
- Spaced repetition can significantly improve long-term retention

6. Automaticity:
- Repeated practice can make skills or recall more automatic, requiring less conscious effort

7. Pattern recognition:

- *Repetition helps identify underlying patterns and principles*
- *Enhances ability to apply knowledge to new situations*

8. Reinforcement of key concepts:

- *Emphasizes core ideas through multiple exposures*
- *Helps prioritize and retain essential information*

Thank you, buddy, for these brilliant eight points. In "Lorne-speak" it comes down to the following: Repetition is what **practicing** is all about. Actors practice their lines; professional athletes practice their sport much more than live games. Great teachers spend a lot of time lesson planning. The practice or repetitive activities is what creates the idea of outliers as brilliantly illustrated in Malcolm Gladwell's best-selling book *"Outliers - The Story of Success."* Here's how he applied the concept:

1. **Definition:** Gladwell used "outlier" to refer to people who achieve extraordinary success or stand out in remarkable ways from the norm. Think Tiger Woods, Michael Jordan, Beethoven and Emily Dickenson.

2. **Focus:** Instead of viewing outliers as anomalies to be discarded (as in statistics), he saw them as subjects worthy of study to understand the factors behind their success. What did they all have in common? Was there an algorithm somewhere? A menu for their success?

3. **Key argument:** Gladwell posited that extraordinary achievement is less about individual talent and more about circumstances, opportunities, and cultural legacy. What was unique about each one of them?

4. **Examples:** He discussed outliers such as Bill Gates, The Beatles, and successful hockey players, examining the unique circumstances that contributed to their success.

5. **10,000-hour rule:** One of his famous claims was that achieving world-class expertise in any skill requires roughly 10,000 hours of practice. Sure enough, they all had this in common.

6. **Cultural legacy:** He explored how cultural background and historical circumstances play a role in creating outliers.

7. **Opportunity**: Gladwell emphasized how access to unique opportunities often distinguishes outliers from their peers. They all seemed to make their own breaks.

Gladwell's use of "outlier" expanded the term beyond its statistical origins, applying it to successful individuals and exploring the complex factors that lead to their exceptional achievements.

AI is number seven above on steroids for each and every one of us. It has created a truly unique opportunity in time. AI accelerates our knowledge and this will help us make those breaks. The playing field is level because we all have access to it. The amount of time and effort you put into AI i.e., practice...the more powerful your Value Moment is going to be and this translates into success.

Back to my scenario early on in this book about flying an airplane. Pilots have to spend a certain amount of time in a flight simulator before they ever get in the cockpit with a live instructor who can take over the controls if you mess up. In addition, you have to spend a certain amount of time with the instructor by your side in flight before they allow you to fly solo. You should approach using AI the same way. Simulate situations or questions for AI to begin with. Read books like this one and others like the new pilot uses to learn to fly. Hire a certified trained AI business coach. (I know where you can find one, by the way). Then, once you have done all of this, begin your solo AI flight.

To "value and serve" you is my greatest goal with this book. Perhaps I was a lot like many of you when it came to AI. I was curious but not engaged. Then one day, I used my chatbot friend, McClear, for the first time. Initially I found it to be an "aha" moment. Then I started to see tremendous value. It was my "moment of value" when I discovered using AI. But, the more I discovered about AI, the good and the bad, the more I put my thinking cap on. Hopefully by now you are getting the picture as to why I titled this book *"AI – Careful Thinking."*

CHAPTER SIX

PIVOT

In this chapter I discuss the cost of having to pivot. Pivoting is in most cases, caused by having incomplete data. When decisions are made around incomplete data, often you miss the mark and have to change courses. AI is wonderful at giving us much more complete data.

Possibilities have always inspired me. I can remember way back 40 years ago, I had an idea and I asked a tech buddy of mine, "Hey Tony, can this be programmed?" When Tony said, "Yes!"...I was off to the races! Vision has always inspired me and passion always fueled me because when I combined vision and passion, I became passionate! My idea became an overnight phenomenon and I was working with some of the biggest brands in the world.

But, there is a catch.

Even though I had the vision and the passion, the actual value of my idea was limited based on incomplete data. When you have incomplete data, essentially you don't have all the facts. When this happens, you have to "pivot" and when you do...well...75% of new businesses fail within the first 15 years. Ever wonder why?

The catch is where AI really comes into play. Many authors write about the need to be flexible and have agility. The popular word for this is being able to "pivot." To help you understand the problem with the idea of pivoting I asked McClear, "What are the challenges with pivoting all the time?"

McClear said:

Pivoting frequently can definitely be challenging. Here are some of the main issues that people and organizations might face:

1. **Loss of Focus:** *Constantly shifting direction can make it hard to maintain a clear vision or focus. This can dilute efforts and make it difficult to achieve long-term goals.* That's me.

2. **Resource Drain:** *Each pivot often requires a reallocation of resources, whether it's time, money, or human capital. This can lead to inefficiencies and increased costs.* Been there, done that!

3. **Team Morale:** *Frequent changes can be demoralizing for teams. It can create a sense of instability and make it hard for team members to stay motivated and engaged.* I saw this happen too.

4. **Customer Confusion:** *If a company is continually changing its product or service, it can confuse or frustrate customers, who might struggle to understand what the company is offering. Yep, you think you are selling them what they want, and you're wrong.* Not a good feeling.

5. **Brand Identity Issues:** *Repeated pivots can erode a brand's identity. Consistency is important for building a strong brand, and constant changes can undermine brand recognition and trust.* Think, COKE, APPLE, BMW, INTEL, AMAZON and ROLEX.

6. **Learning Curve:** *Each new direction comes with its own learning curve. This can slow down progress as teams need time to understand new markets or technologies.* The Internet junkyard is filled with some pretty good ideas, but the owners of these ideas spent too much time changing directions and simply ran out of money.

7. **Missed Opportunities:** *By constantly changing direction, there's a risk of not fully capitalizing on any single opportunity. It can be challenging to build momentum if you're always starting over.* If you have ever moved, and I sure you have, shutting down a house, packing it up and then moving is a pain. Then once you arrive at the new place, the inverse begins.

8. **Strategic Confusion**: *Constant pivots can create a lack of strategic coherence. It becomes difficult to establish a clear strategic path or to communicate it effectively to stakeholders. Strategies are backed up with tactics that carry out each mission.* When strategies change, tactics that are already in place are hard to stop.

9. **Operational Disruption:** *Frequent changes can disrupt operational processes and workflows, making it harder to maintain efficiency and effectiveness. In a military operation, when troops are deployed and you have radio silence, then changes take place at command head-quarters...how are the troops supposed to know these changes?* Things go haywire pretty fast.

10. **Decision Fatigue:** *Constantly making decisions about new directions can lead to decision fatigue, where the quality of decisions declines over time due to the sheer volume and frequency of choices.* This is where you find yourself saying, "this is making me nuts!"

Really cool...right? If I took credit for this information, I would look really smart! I did add the last sentence to each one of McClear's comments. I couldn't resist. Sorry buddy.

One last comment from McClear: *Finding the right balance between flexibility and agility is crucial. While pivoting can be essential for adapting to new information and market conditions, doing so too frequently without clear purpose can create more problems than it solves.* Thank you, buddy.

What just happened? Value just happened. McClear just served me valuable information from something known as "Generative AI." McClear is a form of Generative AI that helps with content creation and information retrieval. In other words, Generative AI is a broad field of Artificial Intelligence, while McClear is a specific implementation of it.

Generative AI, through the use of what is known as a "large language model" (LLM), uses architecture and algorithms that allows the use of very large models, often with hundreds of billions of parameters. Such large-scale models can process massive amounts of data, often from the Internet, but also from sources such as the Common Crawl, which comprises more than 50 billion web pages, and Wikipedia, which has approximately 57 million pages. Much of this data was collected by the three billion smartphones all over the planet. That's right, you might have even contributed to this answer on pivoting and didn't even realize it!

So, with the simple question about pivoting, I was able to get a precise and accurate answer that adds unparalleled value to the impact that pivoting has on any business.

The word PIVOT in many ways has become synonymous with "it's ok to change your mind," and this is reflected in politics, religion, and education throughout the world. In politics the popular term for pivoting is called a "flip-flopper." Often, politicians are accused of being a flip-flopper because they made claims based upon their own opinions rather than the actual hard data or facts that support a political opinion, or maybe they just changed their minds. It is called "the truth." The truth is the only value that in the end will serve any of us well.

A change in posture relative to how you use technology is important. Many of us hesitate to use AI because we are afraid of it or think it is going to somehow take over our minds. Consider the following analogies below.

Fire will burn you if you don't use it properly. Electricity will shock you and water can drown you. The key is finding the place where technology (in this case, AI) serves you well. Some say that AI is the biggest thing to impact human beings since electricity. In my case, AI can be compared to when humans discovered the wheel. Let me explain.

Before the wheel mankind had fire, hammers, spears, knives, shovels and all kinds of tools to make their lives easier. In my life I have had radios, televisions, phones, cell phones, the Internet, email, texting, and social media. All of these helped me function at a very high level. But, like not having the power of a wheel to more easily move things, me not having AI meant that all of my ideas (business ideas in particular) were fundamentally based on very simplified data at best. AI and all of its uses are a game changer for me...and for you by the time you finish this book! But, remember...careful thinking is required.

What used to take an army of experts (Harvard, Wharton and MIT MBA grads) to do market research and create reports, now takes seconds using AI. This means that people who think they have good ideas no longer have to rely on market trends from a sample of one. It is amazing how many great ideas never get off the ground because the seed of the idea was really good but where they planted the seed and how they took care of it was wrong and it never bore any fruit. Now along comes AI and wouldn't you know it...a simple trip down Generative AI lane and a billion data points later, you have the right way to plant that seed!

Back to pivoting. Here is the bottom line. Although many business guru's talk about the importance of knowing how and when to pivot, the truth of the matter is that if you have the right data, you don't need to pivot. To me, pivot is really a cop out to the entrepreneur and the marketer.

I remember as a kid believing that if you tied a string to a tin can 100 feet long and then tied a tin can to the other end you had a telephone. It was imaginary. Why wouldn't it work, after all, the phone in our home was just connected to wires that were connected to telephone poles? What we didn't realize is that the inner workings of the phones in the house and what was in the wires that enabled our voices to travel thousands of miles was a bit more complex than the string and the tin cans.

With AI the idea of needing to pivot is changing. We can now afford a team of experts, they are just virtual. We no longer need to be "tin can telephone" people operating on our own whims. AI is much more than an "aha" moment, it is our Value Moment.

CHAPTER SEVEN

LEARNING TO SWIM

> **In this chapter I refer to the saying, "you can't learn to swim reading a book." Although I am happy you are reading this book, when it comes to AI, you need to learn and the best way to learn is to use it. Like swimming, learn to tread water, then dog paddle and eventually you will be able to swim!**

One of my favorite sayings is that you cannot learn to swim by reading a book. Likewise, you cannot learn to use AI if you don't practice using it. Learning to swim is a really important activity, especially if you live near water. The water in our lives is technology because we live in the information age.

The United States has gone from farming, to industrial to the technology/information age in less than 250 years. We are a society based in capitalism and wealth is generated by having access to the most accurate information. When you look at the dramatic shift in wealth in the US alone in the last four decades, information has been the common theme throughout.

Let's ask McClear and see what he has to say about wealth creation in the last 50 years in the US.

Here is his response: *To answer this question accurately, we'd need to analyze economic data from the past 50 years. However, I can provide a general overview of industries that have been significant wealth creators in the US since the 1970s, based on historical trends and major economic shifts:*

1. Technology:
- Software, hardware, Internet companies
- Companies like Apple, Microsoft, Google, Amazon, Facebook

2. Finance:
- Investment banking, hedge funds, private equity
- Financial services and insurance

3. Healthcare:
- Pharmaceuticals, biotechnology, medical devices
- Health insurance and healthcare services

4. Energy:
- Oil and gas companies
- More recently, renewable energy sectors

5. Real Estate:
- Both commercial and residential sectors
- Real Estate Investment Trusts (REITs)

6. Retail:
- Large-scale retailers like Walmart
- E-commerce (overlapping with technology)

7 Entertainment and Media:
- Film, television, streaming services
- Gaming industry

8 Telecommunications:
- Mobile networks and Internet service providers

9. Aerospace and Defense:
- Aircraft manufacturers, defense contractors

10. Automotive:
- Traditional car manufacturers
- More recently, electric vehicle companies

It's important to note that wealth creation can be measured in different ways (e.g., market capitalization, GDP contribution, job creation) and the ranking might vary *depending on the specific metric used. Additionally, some industries have seen more growth in recent decades than others.* Wow, that's a mouthful McClear!

Number one on the list is technology! In addition to this, technology is the main driver behind the growth of the other nine on the list. The one thing all of these industries did was learn to swim in the technology age. Technology has created *"accelerated intelligence."*

Clearly by this point in the book you can see that I am laying the groundwork for you to get excited about AI and embrace it as a helpful tool. My goal is to help you find your Value Moment using AI.

For many of us, we think AI is only for the brainiacs or those big companies who can afford the Harvard and MIT MBA graduates. This is simply not the case. In fact, the one really significant aspect of AI is that it truly does level the playing field for all of us. Using AI, you can access the same information in the same period of time as the brightest Harvard MBA that ever walked the halls of that hollowed university! You just don't need to be intimidated any longer.

About 40 years ago I had a buddy that I used to play basketball with at the YMCA. He was an insurance broker and his wife was in the advertising business. One evening we had a dinner party at my house and they were in attendance. The wife showed up late because her plane was a bit delayed from Seattle to Portland, Oregon where I lived. When she arrived, she was all excited about a new client she got that day. The client was M I C R O S O F T. That's right, she even had trouble saying it back then. It was more like she said the name of the company is mic-ro-soft. When she started talking about operating systems for these new things we all just got called desktop computers...we were not impressed since our IBM machines already had some kind of engine inside them that seemed to work. Well, you can probably guess the rest of the story. She is enormously wealthy today!

To add fuel to this fire, a year or two later I had my own offices in downtown Portland doing something (I think it was an entrepreneurial education company) and the guy in the office next to mine (a guy named Curt is all I can remember) seemed to be doing very well. There were times that we would cross paths. Turns out, Curt was in sales for Microsoft. I remember him telling me that the demand for this new operating system software was incredible and that I should think about joining him. I still didn't get it. Curt retired very young.

The advertising lady and my buddy Curt, learned to swim...and they were swimming downstream while the rest of us continued to swim upstream. Throughout this book I have referred to changing your posture on AI. Changing your posture is all about taking the time to learn to swim and making sure you are not swimming upstream. But, make sure you know how to swim.

Like swimming where there are many different types of swimming strokes, AI has many different applications as well. Look at it this way with a swimming analogy compared to AI. We are going to learn to tread water then dog paddle, then breaststroke, free style, and some of us will even learn to backstroke and do the butterfly!

The point is this, you are going to be in the AI water whether you like it or not. Using SIRI or ALEXA is like treading water. Still that's ok. The exciting news is that once you begin to learn to dog paddle and beyond, you are going to be opening new doors to opportunities that will change your life as you begin to master AI.

Look at it this way, if you are reading this book, you are at least at the shore of the lake first. That means you got a good parking spot and the best place on the beach. The real fun though is in the water! It is time to swim!

CHAPTER EIGHT

CONFRONT YOUR OWN INCOMPETENCE

> **Ouch! Incompetence is a word that many of us don't like to admit. The reality is that often we are incompetent and we make mistakes. I challenge you in this chapter to see AI like having a really, really smart person by your side helping you make better decisions because they have access to knowledge that you don't have. Not a bad idea, right?**

™

Wow, Lorne that sounds harsh. For most of us, if we are honest, we need to face this true statement. As a lifelong entrepreneur, this statement hits home hard. It is amazing how many people I have met over the years that almost made it big! When you ask them (including myself) why did your idea not work out...regardless of the excuses, it was a form of incompetence. Here is what McClear has to say about incompetence:

Incompetence refers to the lack of ability, skill, or knowledge to perform a task or fulfill a role adequately. It describes a state of being incapable or unqualified to handle certain responsibilities or duties effectively.

Key aspects of incompetence include:

1. Insufficient knowledge or expertise in a particular area

2. Inability to meet the required standards or expectations of a job or task

3. Poor decision-making or judgment in relevant situations

4. Lack of necessary skills or experience to perform effectively

Incompetence can manifest in various contexts, such as professional settings, personal responsibilities, or specific skill-based activities. It's important to note that incompetence is often situational and doesn't necessarily reflect on a person's overall capabilities or worth.

Here is what I have to say about incompetence. In the middle of the word you find the word "compete." In a free enterprise world, our ability to compete and win is the goal. You want to close the deal.

The etymology of the word incompetence is the following: incompetent (adj.)

1610s, "insufficient," from French incompétent, from Late Latin incompetentem (nominative incompetens) "insufficient," from in- "not" (see in- (1)) + Latin competentem (see competent). Sense of "lacking qualification or ability" first recorded in the 1630s.

If I am incompetent, it actually means that I lack the qualification or ability to compete. Even the great Michael Jordan, arguably the greatest basketball player in history, was lacking skills when it came to baseball when he tried to switch sports. He was very competent in basketball, and incompetent in baseball. He was still awesome. He just wasn't competent in baseball like he was in basketball. He clearly could not compete at the same level in baseball. AI increases your competency immediately because you have access to information and knowledge that you didn't previously have. That is a game-changer.

Now, here is what McClear has to say about how AI can increase a person's competence:

AI can increase a person's competence in several ways:
1. Personalized learning: *AI-powered educational platforms can adapt to individual learning styles and paces, providing tailored content and exercises to enhance knowledge acquisition and skill development.*

2. ***Augmented decision-making:*** *AI tools can analyze vast amounts of data and provide insights, helping individuals make more informed decisions in their personal and professional lives.*

3. ***Skill enhancement:*** *AI-powered tools and simulations can offer practice environments for various skills, from language learning to complex problem-solving, allowing users to improve their abilities in a safe, controlled setting.*

4. ***Information access and organization:*** *AI-driven search engines and knowledge management systems can help individuals quickly find and organize relevant information, enhancing their ability to stay informed and up-to-date in their field.*

5. ***Task automation:*** *By automating routine or time-consuming tasks, AI frees up time for individuals to focus on higher-level thinking and skill development.*

6. ***Real-time feedback and coaching:*** *AI systems can provide immediate feedback on performance, helping individuals identify areas for improvement and offering suggestions for enhancement.*

7. ***Collaborative enhancement:*** *AI can facilitate better teamwork by optimizing communication, task allocation, and project management, thereby improving overall team competence.*

8. ***Cognitive augmentation:*** *AI assistants can serve as external memory aids, helping individuals recall information and manage complex tasks more effectively.*

9. ***Predictive analysis:*** *In fields like finance or healthcare, AI can help professionals make more accurate predictions and assessments, enhancing their overall competence.*

10. **Continuous learning support:** *AI can identify knowledge gaps and suggest relevant learning resources, supporting lifelong learning and professional development.*

Here is the truth, competence is directly related to experience from learning about anything. Years ago I was doing a project in the oil fields of North Dakota and working with a salty old oil guy from one of the huge oil companies. He was talking about the dangers of working on an oil rig and how quickly someone can lose their life. What I found interesting was the fact that most, if not all of his examples on how to do something, were based on real life experiences he had to deal with. Much of it was frightening. But the truth in his voice was real. I will never forget it when he stated, those who are competent live and those who are not either get killed or lose a limb.

I believe that as you find your Value Moment using AI, one of the most valuable aspects will be the overnight increase in competence because AI uses the experience from billions of pieces of data. If you were going to be working on an oil rig, imagine how valuable it would be to get immediate knowledge from many people like I did from the guy about the oil rig?

In business and in life, incompetence is our worst enemy. The world is filled with people who have used the statement, *"if I would have only known."* AI can offer you the *"known"* information that you wish you would have known about prior to making bad decisions in life. It simply helps you make more informed decisions and lower or eliminate your incompetence.

Truth be known, we are literally born incompetent. The famous business and marketing writer, Seth Godin, asks the question, "Why do they call a child a toddler?" His answer is pretty powerful. They call them toddlers because when they first learn to walk, they toddle back and forth. They take a step or two, wobble, fall down and then get back up and try again until they learn to walk.

You could say that in the beginning, they are incompetent when it comes to walking. But, to overcome the incompetence, they begin to learn from the data their minds receive each time they fall. In time, all of that data is compiled and their little brains begin to believe they can walk…and they do.

Your journey to understanding how AI can impact your incompetence will begin with your very first step. Undoubtedly you are going to fall down in the beginning. My hope is this book helps you understand the fact that learning to use AI (and what it can do for you relative to your own incompetence) will become priceless. And, like the toddler who eventually learns to walk, you will be able to go places you never dreamed of while you were toddling in your own incompetence. Again, I know this whole "incompetence" word sounds a bit harsh but if you think about it, learning from our incompetence is priceless.

AI in reality takes you from being incompetent to being more-competent almost immediately. MBA and PhD's spend years and hundreds of thousands of dollars to acquire competence so they can teach or consult with those (I will say this in a kinder, gentler way) "less competent." AI is truly a game-changer.

One last thing on incompetence. No matter where you are in life, you will find people much smarter than you using AI. Why? Because there is something they, too, simply don't know. One characteristic I have discovered from really smart people is that they are not afraid to admit it when they are incompetent about something. The difference is they are always willing to learn. AI is all about learning. Make it your friend and find your Value Moment. And always remember, careful thinking is critical to your success using AI.

CHAPTER NINE

TIME TO LEARN THE BASICS OF AI

Depending on who you talk to or what article you last read, what AI is, is all over the map. In this chapter I will clear up these questions and give you more perspective to help you carefully see the value that AI can be to you.

TM

As the ring announcer for major boxing matches says, "LET'S GET READY TO RUMBLE!" For starters, let's dispel the rumors of what AI is and is not.

When addressing concerns about AI safety, it's important to approach the topic thoughtfully and avoid dismissing all concerns outright. Here's a balanced perspective on some common worries about AI:

1. **Job displacement:** While AI will likely automate some tasks, it's also expected to create new job opportunities and enhance human productivity in many fields. The key is adaptation and reskilling. Treat AI the way a golfer does a new driver. New technology in shafts and clubhead design and materials allows us to hit the ball farther and straighter. If my competition is using that technology and outdriving me, maybe it's time for me to adapt and get a new golf club.

2. **Privacy concerns:** As with any technology handling personal data, proper regulations, and ethical guidelines are crucial to protect individual privacy. Stay alert and thoughtful.

3. **Bias and fairness:** AI systems can perpetuate existing biases if not carefully designed. Ongoing research and diverse teams are working to address this issue. In the meantime, be discerning.

4. **Safety and control:** Researchers are actively working on AI alignment to ensure AI systems behave in ways aligned with human values and intentions. Pay attention!

5. **Existential risk:** While some worry about super-intelligent AI posing risks, many experts believe with proper development and safeguards, advanced AI can be a great benefit to humanity. I like the saying, keep your friends close and your enemies closer. AI is not your enemy but, staying dialed in to AI is your best safeguard.

6. **Misinformation:** AI can be used to create convincing fake content, especially deep fakes, but it's also being developed to detect such misinformation. Unless you have been under a rock in the last 10 years with all the fake news and content on social media and news outlets, you know that testing what you hear is critical.

7. **Rather than viewing AI as inherently dangerous**, it's more productive to focus on responsible development, appropriate regulation, and leveraging AI's potential benefits while mitigating risks. Many researchers, ethicists, and policymakers are working to ensure AI development proceeds safely and ethically. That is good news by the way.

8. **Artificial Intelligence is a branch of computer science** that aims to create intelligent machines that can perform tasks that typically require human intelligence. These tasks include learning, problem-solving, perception, and language understanding.

Instead of me taking a bunch of time to create a list, I asked McClear to help me collaborate on the key concepts and applications of AI. Here is what we came up with:

Key Concepts

Machine Learning: A subset of AI that focuses on the development of algorithms and statistical models that enable computer systems to improve their performance on a specific task through experience. In Lorne-speak the machines or computers, look for patterns. The best example would be Netflix. The computers at Netflix keep track of what you watched recently and based on the pattern of what you watched, they can learn to make recommendations of similar programs that might be interesting to you and present them to you. These machines can do the same for much more than your Netflix interests but the same principle applies. They look for patterns in pretty much every topic or subject you can think of and present them to you along with other suggestions for your consideration. So for example, let's say you are visiting a new city and you asked someone who knows you where to stay? They might give you suggestions of where to stay because they know you like nicer hotels. At the same time, they might suggest restaurants and local events that you might be interested in as well. They know you and based on the knowledge of you, or the patterns of interest you have demonstrated to them, they can make a more informed suggestion(s).

Deep Learning: *A more complex form of machine learning that uses artificial neural networks inspired by the human brain to process data and make decisions. In Lorne-speak, deep learning differs from machine learning like this. With machine learning when you visit a new city, you get suggestions from one person. With deep learning you get suggestions from hundreds of people who know you.*

Natural Language Processing (NLP): *The ability of machines to understand, interpret, and generate human language. In Lorne-speak NLP is pretty easy to explain. Most likely you have experienced this already if you have ever used SIRI or ALEXA. It's like the computer (chatbot) is a student, and the data is its textbook. The more it studies and practices, the better it gets at understanding and using language. In its simplest form, if you tell the chatbot* that an apple is a fruit over and over and then someone asks the chatbot if an apple is a fruit, the computer follows the pattern of knowing that yes, an apple is fruit. If you tell the computer that speaks through the chatbot a billion suggestions of what things are, it can quickly see the patterns and provide the answers almost instantaneously.

Computer Vision: The field of AI that trains computers to interpret and understand visual information from the world. In Lorne-speak, computer vision is like teaching a computer to see and understand pictures, just like you do with your eyes and brain. The computer looks at lots of images and learns to recognize different things, like iconic landmarks, cars, or food. It does this by breaking down each picture into tiny pieces called pixels and looking for patterns. The computer practices with many examples until it can spot these patterns quickly. When the computer, for example, in your phone sees a new picture, it can tell you what's in it, almost like magic! This helps the computer do cool things like unlocking phones with your face or helping self-driving cars avoid obstacles on the road.

Robotics: The branch of AI that deals with the design, construction, and operation of robots. This one is going to be more of a challenge in Lorne-speak. For starters, a robot is just a hunk of metal without the computer inside of it giving it commands. It is a machine. Once a human programs the machine, using machine learning, NLP, deep learning and computer vision, the robot is then mechanically activated to move. I had my shoulder replaced a few years ago and the surgeon used a robot to more accurately cut the bones in my shoulder than he could do with his bare hands. I was nervous at first until he told me how much more accurate the robot could be than him. My shoulder works perfectly by the way!

Applications of AI (here is a short list of just a few of the applications that McClear provided for me regarding the applications. Pretty impressive.)

AI has numerous applications across various industries:

- *Healthcare: Diagnosis assistance, drug discovery, personalized treatment plans*
- *Finance: Fraud detection, algorithmic trading, credit scoring*
- *Transportation: Self-driving cars, traffic management, logistics optimization*
- *Education: Personalized learning, automated grading, intelligent tutoring systems*
- *Entertainment: Recommendation systems, game AI, content creation*

Challenges and Ethical Considerations

While AI offers many benefits, it also presents challenges and ethical concerns:

- *Privacy and data security*
- *Bias and fairness in AI systems*
- *Job displacement due to automation*
- *The need for AI transparency and explainability*
- *Long-term implications of artificial general intelligence (AGI)*

Artificial Intelligence is no longer just for big tech companies. Small businesses can leverage AI to improve efficiency, cut costs, and enhance customer experiences. Below is a list of what I believe are some of the most obvious uses of AI that we already are seeing and will expand.

1. Customer Service

- **AI Chatbots:** Provide 24/7 customer support, answering common questions and freeing up staff for complex issues.
- **Sentiment Analysis:** Analyze customer feedback to understand satisfaction levels and areas for improvement.

2. Marketing and Sales
- **Personalized Marketing:** Use AI to analyze customer data and create targeted marketing campaigns.
- **Lead Scoring:** Identify the most promising leads to focus sales efforts efficiently.
- **Social Media Management:** AI tools can help schedule posts, analyze engagement, and identify trending topics.

3. Operations and Efficiency
- **Inventory Management:** Predict demand and optimize stock levels to reduce waste and stock-outs.
- **Task Automation:** Automate repetitive tasks like data entry, appointment scheduling, and invoice processing.
- **Financial Forecasting:** Use AI to analyze financial data and predict future trends, aiding in budgeting and planning.

4. Human Resources
- **Resume Screening:** Quickly sort through job applications to identify the most qualified candidates.
- **Employee Training:** Personalized learning experiences based on individual employee needs and performance.

5. Data Analysis and Decision Making
- **Business Intelligence:** AI can analyze large amounts of data to provide actionable insights for decision-making.
- **Predictive Analytics:** Forecast trends, customer behavior, and potential risks.

6. Cybersecurity
- **Threat Detection:** AI systems can monitor networks for unusual activity and potential security breaches.
- **Fraud Prevention:** Identify and prevent fraudulent transactions more effectively than traditional methods.

By strategically implementing AI solutions, small businesses can compete more effectively, improve customer satisfaction, and drive growth in today's digital economy.

As a beginner looking to understand the basics of AI, here are some key concepts you should familiarize yourself with. Most of these we have already discussed earlier in the book. Still, if you are going to use AI, it is a good idea to know what these are and what they have to do with AI.

1. **Machine Learning:** *The core of modern AI, where algorithms learn patterns from data.*

2. **Neural Networks:** *A type of machine learning model inspired by the human brain.*

3. **Deep Learning:** *Advanced neural networks with multiple layers, capable of learning complex patterns.*

4. **Supervised Learning:** *Training models using labeled data.*

5. **Unsupervised Learning:** *Finding patterns in unlabeled data.*

6. **Reinforcement Learning:** *Training models through reward-based feedback.*

7. **Natural Language Processing (NLP):** *Enabling machines to understand and generate human language.*

8. **Computer Vision:** *Allowing machines to interpret and analyze visual information.*

9. **Data Preprocessing:** *Cleaning and preparing data for use in AI models.*

10. **Model Evaluation:** *Assessing the performance of AI models using metrics and testing.*

11. **Ethics in AI:** *Understanding the societal impacts and ethical considerations of AI technology.*

Here are some keywords that make up the AI vocabulary I asked McClear to put together that beginners should be familiar with:

1. **Algorithm:** *A set of instructions or rules for solving a problem or performing a task.*

2. **Dataset:** *A collection of data used to train and test AI models.*

3. **Feature:** *An individual measurable property or characteristic of a phenomenon being observed.*

4. **Label:** *The target variable or outcome in supervised learning tasks.*

5. **Training:** *The process of teaching an AI model using data.*

6. **Inference:** *Using a trained model to make predictions on new data.*

7. **Overfitting:** *When a model learns the training data too well, including noise and outliers.*

8. **Underfitting:** *When a model is too simple to capture the underlying patterns in the data.*

9. **Hyperparameters:** *Configuration settings for AI models that are set before training begins.*

10. **Bias:** *Systematic errors in AI systems, often reflecting human prejudices.*

11. **Generalization:** *A model's ability to perform well on unseen data.*

12. **Gradient Descent:** *An optimization algorithm used to minimize the error in predictive models.*

13. **Epoch:** *One complete pass through the entire training dataset during model training.*

14. **Activation Function:** *A function that determines the output of a neural network node.*

15. **Backpropagation:** *The primary algorithm for training neural networks.*

Thanks to McClear for putting these together. Even as I write this book, you can see how valuable McClear is to me in terms of accuracy and saving time assembling the list. He literally put this all together for me faster than I could type in the question.

These terms form a crucial part of the AI vocabulary. Throughout this book you are going to be introduced to these terms so don't lose any sleep at this point if you don't remember them right now. You are not even in the toddler phase of your learning. As you get started with using AI, you will get in the groove so to speak. And yes, the chances of you ever really having to know what number 15 above is (backpropagation) is remote.

Understanding what AI is not can be just as important as knowing what it is. Here are some key points about what AI is not:

1. **Not human-like general intelligence:** AI systems are designed for specific tasks and lack the general intelligence and adaptability of humans. McClear is my buddy, but he is like a rock sitting on the side of a riverbed unless someone picks it up and throws it in the water.

2. **Not conscious or self-aware:** Current AI does not possess consciousness or self-awareness like humans do. Even when McClear does respond to my command, he is only responding based on data he can access. I can't ask him how he likes this book like I can ask you how you like it.

3. **Not infallible:** AI systems can make mistakes, exhibit biases, and produce incorrect results. Sorry McClear. Yes, you do make mistakes. Why? Because the data you pull your answers from can sometimes be incorrect.

4. **Not magic or omniscient:** AI has limitations and operates based on the data and algorithms it's given. Don't tell this to McClear...it hurts the feelings that he doesn't have!

5. **Not autonomous decision-makers:** Most AI systems require human oversight and don't make independent decisions. This is one of the main topics that are discussed in global ethics panels regarding the use of autonomous weapons as an example.

6. **Not emotional:** AI doesn't have feelings or emotional intelligence in the way humans do. Even though as a writer I have fun with the illusion that McClear is my buddy, he is not a he or a she, rather a computer program. Sorry buddy.

7. **Not creative in the human sense:** While AI can generate novel outputs, it doesn't have true creativity or original thought. Original thought is probably the main thing to remember here. AI can only provide output based on human thought that their programming pulls data from.

8. **Not a replacement for human expertise:** AI is a tool to augment human capabilities, not replace human judgment entirely. It is up to us as the users of AI to make the judgment on the information or results we receive from AI.

9. **Not always the best solution:** Sometimes simpler, non-AI approaches are more appropriate for certain problems. I don't know though, McClear is pretty handy!

10. **Not inherently ethical or unethical:** The ethics of AI depend on how it's developed and used by humans. THIS IS HUGE. CAREFUL THINKING REQUIRED!

11. **Not improving without human intervention:** AI systems generally don't learn or improve on their own after deployment without being retrained. I say this however with a certain amount of reservation. Some experts are already sending out warning signals.

12. **Not able to understand context like humans:** AI often struggles with nuance and context that humans grasp intuitively.

Understanding these limitations helps set realistic expectations about AI's capabilities and applications. Let's face it, I have covered a great deal of information in this book. For me, to find my Value Moment with AI, I felt it was fundamentally important to get the entire picture of AI, at least to date.

AI is fundamentally about creating systems that can perform tasks typically requiring human intelligence. The only difference is that AI can perform many of these tasks much faster with much more detail. Below is a list of important functions that AI does:

1. **Problem-solving**: AI systems are designed to tackle complex problems and make decisions based on data and algorithms. The keyword here is the word, complex. Complex is a relative term. A beauty salon may consider it complex to figure out how to sell more products to their clientele. A micro-brewery might consider it complex to adjust their menu to sell more IPA beer. Problems are problems and AI can be a valuable tool to solve them.

2. **Learning and adaptation:** Many AI systems can improve their performance over time through experience, a process known as machine learning. So the brewery and the salon can begin to input their own data and utilize prompt engineering along with data that AI is able to access from other breweries and salons around the world and learn from these in order to provide more accurate results.

3. **Pattern recognition:** AI excels at identifying patterns in large datasets, which can be applied to various fields like image recognition, natural language processing, and predictive analytics. As mentioned in number 2 above, AI will look for these patterns. That's what McClear does for me!

4. **Automation:** AI aims to automate tasks that previously required human intervention, increasing efficiency and productivity. McClear is constantly reminding me that he can write the next book without me...good luck buddy!

5. **Mimicking human cognition:** While not necessarily replicating human thought processes, AI attempts to achieve human-like results in specific domains. This is a

bit frightening but is actually happening. Kind of cool, especially when you think of producing promotional videos using virtual actors who are significantly less expensive than live actors.

6. **Data processing:** AI systems can analyze and interpret vast amounts of data much faster than humans, leading to insights and predictions. This is just smart. The data is out there that can help us, why not correlate it into an effective use?

7. **Interdisciplinary field:** AI combines computer science, mathematics, psychology, linguistics, and other disciplines to create intelligent systems. The goal here is to provide results that take these disciplines into consideration to provide more accurate outcomes using AI.

8. **Continual evolution:** The field of AI is constantly advancing, with new techniques and applications emerging regularly. The smarter we get as humans, the smarter AI can become. Again, in many areas this is really powerful and effective. In the wrong areas it can be very dangerous. Ethics warning here.

9. **Ethical considerations:** As AI becomes more prevalent, questions about its impact on society, privacy, and decision-making processes are increasingly important. Amen.

Here's a summary on how a novice should begin using AI in Lorne-speak:

1. **Start with user-friendly AI tools:**
   - AI-powered writing assistants (e.g., Grammarly)
   - Image generation tools (e.g., DALL-E, Midjourney)
   - Voice assistants (e.g., Siri, Google Assistant and Alexa)

2. **Explore AI-enhanced applications:**
   - Photo editing software with AI features
   - AI-powered music creation tools

- Smart home devices

3. Experiment with chatbots:

- Use conversational AI like ChatGPT4.o or Claude 3.5 for various tasks (I renamed Claude 3.5. His name is McClear)
- Learn prompt engineering to get better results

4. Leverage AI in everyday tasks:

- Use AI-powered features in productivity tools (e.g., smart compose in Gmail)
- Try AI-enhanced search engines (Use Google and learn about these)

5. Join AI communities:

- Participate in forums discussing AI applications
- Attend webinars or workshops on practical AI use

This approach focuses on hands-on experience with accessible AI tools. In the remaining chapters of this book you are going to learn the nuts and bolts of AI. You don't need to become an expert overnight. This book will open your eyes to the majority of the AI world, what it can do, why you should use it, and how you can maximize its use. It's all about finding that Value Moment relative to you.

CHAPTER TEN

CHATBOTS - WHERE THE FUN BEGINS!

In this chapter I explain how incredible chatbots are but want you to keep in mind, they do make mistakes. We have leaned on technology for long enough now that we tend to just take it for granted we are getting accurate information. With chatbots, careful thinking is paramount.

Ok McClear...it is your time to shine! We are going to talk about you in this chapter. A chatbot is like a computer friend you can talk to by typing messages. Some can even talk to you! It's a really cool program that understands what you write or say and gives you answers, almost like having a conversation. You can ask it questions, get help finding information, or even play games with it. Chatbots use artificial intelligence to learn and get better at talking to people over time. While they can be very helpful and fun, it's important to remember that they're just clever computer programs, not real people. Sorry McClear (:

A chatbot is a computer program designed to simulate human conversation through text or voice interactions. Chatbots take Siri and Alexa to the next level. Personally, the first time I used one, I was blown away. A great deal of this book, in fact, was made possible because the chatbots make research both accurate and fast. Half of the battle of writing a book is finding data to back up your written thoughts. Just sayin...to all of you authors out there!

Here are the key points about chatbots from my chatbot buddy, McClear:

1. Purpose: *Chatbots are created to interact with users, answer questions, and sometimes perform tasks.* If chatbots had a mind of their own, I am certain that McClear would give me a mouthful!

2. Technology: *They use natural language processing (NLP) and artificial intelligence (AI) to understand and respond to user inputs.*

3. Types:
- *Rule-based chatbots: Follow pre-programmed rules.*
- *AI-powered chatbots: Use machine learning to improve responses over time.*

4. Applications:
- *Customer service*
- *Information retrieval*
- *Personal assistants (like Siri or Alexa)*
- *Entertainment*

5. Interfaces: *Chatbots can be integrated into websites, messaging apps, or voice assistants.*

6. Limitations: *While advanced, chatbots may struggle with complex queries or understanding context fully.*

Chatbots aim to provide quick, efficient, and sometimes personalized interactions, though their capabilities vary widely depending on their design and purpose.

The future of chatbots is likely to be dynamic and transformative. Here are some key trends and developments we can expect:

1. Advanced AI and Natural Language Processing:
- *Chatbots will become more sophisticated in understanding context, nuance, and emotion.*

- *They'll engage in more natural, human-like conversations.*

2. Multimodal Interactions:
- *Future chatbots may integrate text, voice, and visual elements.*
- *They could analyze images or videos alongside text to provide more comprehensive assistance.*

3. Personalization:
- *Chatbots will offer increasingly personalized experiences by learning from past interactions.*
- *They may adapt their communication style to individual users' preferences.*

4. Expanded Use Cases:
- *Beyond customer service, chatbots may play larger roles in healthcare, education, and personal productivity.*
- *They could serve as virtual therapists, tutors, or life coaches.*

5. Integration with IoT and Smart Devices:
- *Chatbots may become central interfaces for controlling smart homes and IoT (Internet of things) devices.*

6. Improved Emotional Intelligence:
- *Future chatbots might better recognize and respond to human emotions.* Nice try McClear...this is a long, long way off buddy...in my opinion never, just saying.

7. Augmented Human Workforce:
- *Rather than replacing humans, chatbots may work alongside them, handling routine tasks and freeing up humans for more complex work.* I think McClear is thinking of running for office.

8. Ethical and Privacy Considerations:
- *As chatbots become more advanced, there will likely be increased focus on ethical AI use and data privacy.*

9. *Blockchain and Decentralized Chatbots:*

- *Some predict the development of decentralized chatbots that operate on blockchain technology, potentially offering more security and transparency.*

- *Blockchain is like a special digital notebook that many people share. When someone wants to add new information, like a trade or transaction, everyone in the network checks if it's correct. If they agree, the new information is added to everyone's copy of the notebook. Once added, this information can't be changed or erased. This system helps keep track of things without needing one person in charge, and makes it very hard for anyone to cheat or change the records.*

10. *Seamless Omnichannel Presence:*

- *Chatbots may provide consistent experiences across multiple platforms and devices.*

It's worth noting that while these trends seem likely based on current trajectories, the field of AI and chatbots is rapidly evolving. New breakthroughs could shift these predictions in unexpected ways. I highly recommend that you begin interacting with your favorite chatbot and ask it to comment on whatever is on your mind.

PRO-TIP: Pay close attention whenever it provides a summary of what it says because that summary will open the doors to even greater considerations on the topic.

Remember, a chatbot is a Google search on steroids basically. Google searches simply deliver relevant information through a list of search results. Chatbots go one step farther. They have been designed to deliver interactive, conversational exchanges and specific tasks. A chatbot engages users in a back and forth dialogue versus the static response from a Google search. Chatbots are programmed to ask follow-up questions, including tailored responses, and can adjust the answers it provides based on the context of the inquiry. It is a conversation flow with follow up "suggestions" that help us (hopefully) dig deeper and find more accurate and meaningful answers. They will also present this information into pretty much any format your request and that is powerful.

As you can see, I interact with McClear as if he is real. He is not, sorry buddy. But, he is very reliable in terms of always being ready when I am.

This being said, careful thinking is required. I often come across the "suggestions" and will respond yes to the suggestion but I will always ask another chatbot (sorry McClear) for a counter-suggestion to get a more complete answer before I decide what I actually want to believe about an AI generated response. It's not your fault McClear, I know you can only respond as well as the data you are given!

Here is an example of two different responses to the question: *"Write me a one paragraph answer to the dangers of AI."* Here is ChatGPT4o and then Claude 3.5:

ChatGPT4o: *The dangers of AI are multifaceted and significant, particularly as the technology becomes more advanced and integrated into various aspects of life. One major concern is the potential for AI to perpetuate and exacerbate existing biases, leading to unfair and discriminatory outcomes in areas such as hiring, law enforcement, and lending. Additionally, the misuse of AI for malicious purposes, such as creating deep fakes or automating cyber-attacks, poses serious security threats. There's also the risk of job displacement as automation and AI systems outperform human capabilities in certain tasks, which can lead to economic instability and social disruption. As AI continues to evolve, addressing these risks requires a balanced approach, incorporating ethical considerations, robust regulations, and ongoing dialogue to ensure the technology is developed and applied responsibly.*

Claude 3.5: *Artificial intelligence presents several potential risks as the technology rapidly advances. These include job displacement due to automation, privacy concerns from data collection and surveillance, algorithmic bias leading to unfair decisions, and the possibility of autonomous weapons. There are also existential risks if highly advanced AI systems become misaligned with human values or interests. Additionally, AI could be used to generate convincing misinformation or deep fakes, further eroding trust in information sources. While AI offers tremendous benefits, careful development and regulation are needed to mitigate these dangers and ensure the technology remains beneficial to humanity.*

The very same question and two similar, but very different responses. Claude 3.5 asked me if I wanted more specifics on each topic and ChatGPT4o did not. I found it interesting that Claude talked about the potential of AI *"being misaligned with human values and interests"* and ChatGPT4o talked about *"economic instability and social disruption."* The same exact question but two completely different interpretations of what is most important. And, admittedly, there is no right or wrong answer here as both are very important. The point is, by asking more than one chatbot the same question, we can get more perspective. If I ask Google and Bing (the two most popular search engines) the results I get back are sponsored ads so that is really no answer at all. Search engines though, kind of set the stage and "trained us so to speak" to believe whatever we see first. Danger Will Robinson.

To finish up on chatbots, they are like driving a sports car with 800hp versus a search engine that is like driving a Prius. Both can go fast enough to hurt you if you crash. But, 800hp can get you to the crash point much faster. Here's my point, if you have never driven a sports car with a ton of horsepower, stepping on the gas makes the tires spin out of control quickly. You can step on the gas of a Prius and it just goes forward...your tires will not spin. Google search in other words, is pretty much harmless. You get your sponsored results, then actual results and then more sponsored results! In fact, some experts are predicting that AI is going to eliminate the need for even Google! Kind of like electric cars making the Prius a thing of the past.

Technology was moving fast even before AI. Now it is moving so fast that technology like Google is becoming an antique. Fighter pilots have technology to enable them to see what is out in front of them because the aircraft literally flies faster than the brain can process information and they would fly past their target before they could ever take a shot. AI is kind of the same. So, learn how to use it, how to control it for you, and use your careful thinking button.

CHAPTER ELEVEN

THE MAJOR PLAYERS IN AI TODAY

> **Keeping up to speed on who the major players are in AI and following them on social media and the news will give you a better barometer on the things you need to be aware of. All of them are in a full on sprint regarding AI because most are public companies and their shareholders are nervous.**

The AI landscape is rapidly evolving, with several major players making significant contributions. As of my last update in April 2024, here are some of the key companies and organizations in the AI field: (spoiler alert...McClear helped me out here)

1. Tech Giants:
- *Google (Alphabet): Known for DeepMind, Google Brain, and various AI applications like Gemini, formerly known as Bard*
- *Microsoft: Significant investments in OpenAI (exceeding $10B), Azure AI, and other AI technologies*
- *Amazon: Amazon Web Services (AWS) AI services and Alexa*
- *Apple: Siri and various AI features in its products*
- *Meta (Facebook): AI research in areas like computer vision and natural language processing*

2. Specialized AI Companies:
- *OpenAI: Known for GPT models and DALL-E*
- *Anthropic: Developing AI systems like Claude with a focus on safety and ethics*
- *DeepMind (owned by Google): Breakthrough*

achievements in areas like AlphaGo and protein folding

3. Chinese Tech Companies:
- *Baidu: China's leading search engine, investing heavily in AI*
- *Alibaba: Cloud computing and AI services*
- *Tencent: AI in social media and gaming*
- *TikTok: Using AI extensively to propose items of interest to users*

4. Enterprise AI:
- *IBM: Watson and enterprise AI solutions*
- *NVIDIA: AI hardware and software platforms*
- *Salesforce: AI integration in customer relationship management*

5. AI Chip Manufacturers:
- *NVIDIA: Leading provider of GPUs for AI processing*
- *Intel: Developing specialized AI chips*
- *AMD: Competing in the AI chip market*

6. Startups and Emerging Players:
- *Many startups are making waves in specific AI domains, though the landscape changes rapidly*

7. Research Institutions:
- *Universities like MIT, Stanford, and Carnegie Mellon*
- *Research labs like Allen Institute for AI*

8. Open-Source Communities:
- *Organizations like Hugging Face, fostering open-source AI development*

9. Coaching Companies and Communities
- *Like Fowler International Academy, Global AI University, and Global AI Directory*

It's important to note that the AI field is highly dynamic, with new players emerging and existing ones evolving their strategies continuously, and at an accelerated pace every day.

The competitive landscape and key players can change quickly with technological breakthroughs or strategic shifts.

Understanding the major players in AI is important for several reasons:

1. **Technological landscape**: It helps you grasp the current state of AI technology and its potential future directions along with the speed of change. Different companies and organizations often focus on distinct aspects of AI development. Many of the popular books on AI focus on different aspects as well.

2. **Economic impact:** AI is reshaping numerous industries. Knowing the key players helps in understanding market dynamics, investment trends, and potential economic shifts. Use AI frequently to keep updated on the dynamics, trends and shifts. Just ask!

3. **Ethical considerations:** Major AI players often set precedents in AI ethics and governance. Their decisions can influence industry standards and public policy. This is evolving on a regular basis. Pay close attention to ethics always when it comes to AI usage.

4. **Research and innovation:** Tracking major players allows you to stay informed about cutting-edge research and breakthrough innovations in the field. A simple prompt weekly asking about breakthroughs in AI is a good best practice to develop.

5. **Career opportunities**: For those interested in working in AI, knowing the major companies and institutions can guide career planning and skill development. Again, ask AI who is hiring and what qualifications are in the highest demand?

6. **Societal implications:** The actions of major AI players can have far-reaching effects on society, from privacy concerns to job market changes. Be aware of these implications.

7. **Competition and collaboration:** Understanding the relationships between major players provides insight into

industry dynamics, partnerships, and competitive strategies. New developments and AI usage could impact your business so pay attention.

8. **Policy and regulation:** Governments and regulatory bodies often engage with major AI players when developing policies. Knowing these players helps in understanding the dialogue around AI governance. Do this if you are so inclined.

I am a big believer in having well-rounded knowledge when it comes to the thought leaders in AI. Better problem-solving: Diverse knowledge allows you to approach issues from multiple angles.

1. **Enhanced creativity:** Combining ideas from different fields often leads to innovative solutions. Your ability to incorporate your own creativity with the machine learning aspects of AI will serve you well.

2. **Improved communication:** Broad knowledge helps you connect with people from various backgrounds. Use AI to keep your conversations more interesting and relevant. Create content from the broad knowledge that AI can provide to show your clients that you care.

3. **Adaptability:** A wide knowledge base makes it easier to adjust to new situations or career changes. If AI impacts your job or business in a negative way, it is probably affecting others in a positive way. Stay out in front of the rolling ball. Adapting is key to the new economy that AI will play a major role.

4. **Critical thinking:** Understanding multiple subjects helps you evaluate information more effectively. Then take critical thinking and apply careful thinking as it all relates to you personally.

5. **Personal growth:** Learning about diverse topics expands your perspective and enriches your life. Again, prompts that ask AI things like; what is it about a certain subject that most people overlook is a smart way to move beyond the obvious.

6. Social and cultural awareness: Broad knowledge fosters understanding and empathy for different cultures and viewpoints. This is a major challenge that many of us have with a global economy. Knowing more about other cultures is both smart and kind. It is easy to be offensive and not even know it.

7. Career advancement: Many roles value individuals with diverse skill sets and knowledge. Let AI be your career advancement sherpa.

8. Lifelong learning: A broad knowledge base makes it easier to continue learning throughout life. Have fun with the fact that AI enables you to tap into areas that you have no idea or understanding about. I try to learn something new every day from AI.

9. Informed decision-making: Understanding various factors allows for more comprehensive analysis in personal and professional choices. Comprehensive is the key word here. AI will help you get the complete picture and that offers you the ability to make better decisions and save time and money.

10. More effective careful thinking because you have access to more data. Getting the entire picture of any issue or concept is always more effective.

If you pay attention to these 10 things in everything you do, the results are going to be much better. When it comes to the major players in AI, you can be certain that they pay close attention to this list as well or some adaptation of the list. It is like the 10 Commandments of AI for me.

CHAPTER TWELVE

COGNITIVE BURDEN RELIEF

This might seem like an odd name for a chapter but it really plays into the key driver behind AI. Relieving our cognitive burden is directly related to easing our "mental workload." As humans technology has conditioned us to find ways to make things faster and easier. This is where being careful and ethics are critical.

AI relieves the cognitive burden in humans. Cognitive burden, also known as cognitive load, refers to the amount of mental effort and memory resources required to process information, perform tasks, or engage in thinking processes. It's essentially the "mental workload" experienced by an individual as they navigate various cognitive demands.

One of the statements you will begin to hear about is; "AI Runs On Big Data." It is important that you understand and take to heart exactly what big data means. Big Data sounds like an ominous concept. Big data refers to extremely large and complex datasets that are difficult to process using traditional data processing methods. AI technology basically took the massive database that you hear about when people talk about "it's in the cloud" and made it all easier for us to access.

These datasets (big data) are characterized by the "Three Vs:"

1. **Volume:** The sheer amount of data being generated and collected. Imagine just the data collected from the three billion smartphones on the planet?

2. **Velocity:** The speed at which new data is being created and needs to be processed. In seconds, computer

technology can process results and deliver them to our smartphones.

3. **Variety**: The diverse types of data, including structured, semi-structured, and unstructured data. Structured data is information in spreadsheets and databases. Semi-structured data would be things like RSS feed and tags. (you might want to ask a chatbot about these). Unstructured data are things like social media posts, images and audio files.

For artificial intelligence (AI), big data is crucial in several ways:

1. **Training data:** AI models, especially machine learning and deep learning systems, require large amounts of data to learn patterns and make accurate predictions. It would be difficult to extrapolate a trend or pattern from a sample of one. Big data provides the volume and variety of information needed to generate possible answers for you to consider.

2. **Real-time processing:** AI systems can analyze and process big data streams in real-time, enabling quick decision-making and adaptive responses in various applications. Think milliseconds or microseconds. It is freaking fast!

3. **Pattern recognition**: The variety and volume of big data allow AI to identify complex patterns and correlations that might be invisible to human analysts using traditional analytics methods. There is no easy way to break this down other than using the chess analogy. World class chess players see moves ahead of their competitors so they can "set them up" for future moves. AI sees much further than the world class chess player.

4. **Improved accuracy:** More data usually leads to more accurate AI models, as they can learn from a wider range of examples and scenarios. However, this is all dependent on the accuracy and reliability of the data that is input.

5. **Scalability:** Big data technologies and AI systems can work together to handle and analyze massive datasets that would be impractical for humans to process manually. In Lorne-speak, AI is a supersonic jet, search engines are single engine airplanes. Both, however, require a pilot or someone to program where the plane is going.

6. **New insights:** The combination of big data and AI can lead to novel discoveries and insights across various fields, from healthcare and finance to marketing and scientific research. This is one of the most exciting aspects of AI where it doesn't take over humanity, it enhances humanity. Again, ethics play a key role here.

In essence, big data provides the raw material that AI systems can process, learn from, and use to generate valuable insights and predictions. This synergy between big data and AI is driving innovation and transforming numerous industries. This transformation will create jobs and eliminate jobs. Valuable insights and predictions will keep us on our toes.

Let's go back to the toddler analogy used earlier in this book. Our brains go through a toddler phase just like our physical bodies. We learn from our mistakes because we remember, for example, as a little kid that if you touch a hot wood stove, we get burned. We got big data when we got burned, right?

As our minds continued to grow and learn, we learned to speak, read, write and do math. We learned how to play with our friends, how to feed ourselves, how to make our own beds and so forth. Learning is the theme. Each of these learning experiences combined, created the big data in our lives.

To continue this line of logic on learning, think back to the day when we were first introduced to the calculator. Prior to the calculator, when you wanted to add something up, you did it by hand on a piece of paper. You learned that 5X5 equaled 25 and 63 divided by 9 equaled 7. Then fractions, algebra, geometry and so on. Then one day in 1967 a company called Texas Instruments introduced the first handheld calculator called the "CalTech" and shortly thereafter, we all had them in our hands. As the calculator evolved, more functionality was added enabling us to program them beyond four-function math, and much more. It relieved the cognitive burden of complex calculations from humans, allowing for more focus on higher-level thinking and problem-solving.

The handheld calculator brought about several significant changes for humans, but arguably the biggest change was democratizing access to rapid and accurate numerical calculations. This had far-reaching effects. I asked my buddy McClear (sorry buddy, I didn't forget about you) to elaborate on these effects and here is what he had to say:

1. Education transformation:
- *Calculators changed how mathematics was taught and learned in schools.*
- *Focus shifted from rote memorization and manual calculation to understanding concepts and problem-solving strategies.*

2. Increased efficiency:
- *Complex calculations that once took minutes or hours could be done in seconds.*
- *This saved time in various fields like engineering, finance, and science.*

3. Reduced human error:
- *Calculators significantly decreased arithmetic mistakes in calculations.*
- *This improved accuracy in fields relying heavily on numerical computations.*

4. Accessibility of advanced mathematics:
- *More complex mathematical operations became accessible to a wider audience.*

- *This democratized access to fields that heavily rely on mathematics.*

5.Cultural shift:
- *Changed the perception of what constitutes mathematical skill.*
- *Emphasized the importance of understanding mathematical concepts over the ability to perform calculations manually.*

6.Professional impact:
- *Transformed workflows in numerous professions, from accounting to scientific research.*
- *Allowed professionals to focus more on analysis and interpretation rather than computation.*

The handheld calculator essentially offloaded the cognitive burden of complex calculations from humans, allowing for more focus on higher-level thinking and problem-solving. This shift in how we approach numerical tasks has had lasting impacts on education, professional practices, and our general relationship with mathematics.

AI reduces the cognitive burden on humans in several ways, similar to how calculators reduced the burden of manual calculations. Here's how AI accomplishes this according to McClear:

1. Automation of repetitive tasks:
- *AI can handle routine, repetitive tasks that would otherwise require human attention and mental effort.*
- *This frees up cognitive resources for more complex, creative, or strategic thinking.*

2. Information processing and analysis:
- *AI can quickly process and analyze vast amounts of data, extracting relevant insights.*
- *This reduces the mental load of sifting through large datasets manually.*

3. **Decision support:**
   - AI systems can provide recommendations or predictions based on complex data analysis.
   - This aids decision-making processes, especially in situations with multiple variables.

4. **Natural language processing:**
   - AI can understand and generate human language, assisting in tasks like translation, summarization, or question-answering.
   - This reduces the cognitive effort required for language-related tasks.

5. **Pattern recognition:**
   - AI excels at identifying patterns in data that might be too subtle or complex for humans to notice easily.
   - This offloads the mental effort of detailed observation and analysis in many fields.

6. **Personalization:**
   - AI can tailor information and experiences to individual preferences, reducing the cognitive load of filtering irrelevant information.

7. **Predictive maintenance:**
   - In industrial settings, AI can predict when equipment needs maintenance, reducing the mental stress of constant monitoring.

8. **Memory augmentation:**
   - AI-powered systems can act as external memory, storing and retrieving information as needed.
   - This reduces the cognitive load of memorizing large amounts of information.

9. **Multitasking assistance:**
   - AI can manage multiple streams of information simultaneously, helping humans focus on one task at a time.

10. *Complex problem-solving:*
- *For intricate problems, AI can break down the problem into smaller, more manageable parts or suggest solution strategies.*

11. *Emotional and social intelligence:*
- *Some AI systems can recognize human emotions or social cues, potentially reducing the cognitive effort required in certain social interactions or customer service scenarios.*

By taking on these various cognitive tasks, AI allows humans to focus their mental energy on higher-level thinking, creativity, emotional intelligence, and other uniquely human capabilities. This shift enables people to work more efficiently and effectively, much like how calculators allowed for a focus on mathematical concepts rather than computation.

In each of these contexts, "burden" maintains its core meaning of something weighty or difficult to bear, but it's applied to different aspects of life, society, or professional fields.

The song "He Ain't Heavy, He's My Brother" provides a very interesting backstory to what AI can be to us relative to the word burden. Those iconic words have symbolized the spirit of Boys Town for decades. But many people don't know how it originated.

Back in 1918, a boy named Howard Loomis was abandoned by his mother at Father Flanagan's Home for Boys, which had opened just a year earlier. Howard had polio and wore heavy leg braces. Walking was difficult for him, especially when he had to go up or down steps.

Soon, several of the Home's older boys were carrying Howard up and down the stairs.

One day, Father Flannigan asked Reuben Granger, one of those older boys, if carrying Howard was hard. Reuben replied, "He ain't heavy, Father... he's my brother."

But the story doesn't end there.

In 1943, Father Flanagan was paging through a copy of Ideal magazine when he saw an image of an older boy carrying a younger boy on his back. The caption read, "He ain't heavy, mister… he's my brother."

Immediately, the priest was reminded of a photo of Reuben carrying Howard at a Boys Town picnic many years before. Father Flanagan wrote to the magazine and requested permission to use the image and quote. The magazine agreed, and Boys Town adopted them both to define its new brand.

Now, many years later, the motto is still the best description of what our boys and girls at Boys Town learn about the importance of caring for each other and having someone care about them.

"He ain't heavy" is relevant beyond Boys Town. At some point in our lives, most of us have needed to be carried by someone, metaphorically speaking. And, at some point, we probably carried somebody else. We're human. We stumble. And we look to each other for help when we do.

When we begin to understand AI and its ability to help us relieve our cognitive burdens, it really becomes our friend who helps to carry us and make our lives better. Remember, we live in the information age. Having the ability to process information faster and more accurately helps us to make better decisions and save time. It is what we choose to do with that time that really matters in the long run.

Using AI can eliminate the "shoulda - woulda - coulda's" in our lives. Like any tool, if you use it for good it will produce good. Likewise, if you use it for bad, well…it produces bad. Consider fire, the wheel, weapons, drugs, the Internet, alcohol and on and on. AI is here to stay, fortunately or unfortunately. It can reduce our cognitive burdens when we use it the right way. It doesn't have to be heavy, it can be your brother…use careful thinking and focus on ethics!

CHAPTER THIRTEEN

ETHICS

Ethics and the Use of AI: Navigating
the Moral Landscape of Artificial Intelligence

> **In this chapter the real 800 lb. gorilla is in the middle of the room. Most, if not all of the fears behind AI, are wrapped up in the ethical use of AI. What AI can do now is relatively harmless compared to where the experts predict it is all headed. The ethical use of AI is critical.**

As artificial intelligence (AI) continues to advance at a rapid pace, it brings with it a host of ethical considerations that society must pay attention to. The integration of AI into various aspects of our lives – from healthcare and finance to criminal justice and education – raises important questions about fairness, transparency, privacy, and the very nature of human-machine interactions.

We all need a bit of a wakeup call when it comes to paying attention to our moral compass and when it comes to major changes that impact society as a whole. AI is one of those landmark developments in the history of mankind.

The Promise and Perils of AI

Clearly, AI technologies offer immense potential benefits for humanity. They can help diagnose diseases more accurately, optimize resource allocation, enhance scientific research, and even assist in addressing global challenges like climate change. However, these same technologies also pose significant risks if not developed and deployed responsibly.

Key Ethical Concerns Relative to Our Moral Compass

1. Bias and Fairness: AI systems are only as unbiased as the data they're trained on and the humans who design them. There's a real risk of perpetuating and amplifying existing societal biases, particularly in areas like hiring, lending, and criminal justice. Remember, AI is not human. It is a machine just like your smartphone is a machine. It is programmed by humans. Humans are not perfect and we make mistakes. These mistakes can create bias and unfairness that can impact us in a negative way. Again, like I have said over and over again, careful thinking is required.

2. **Privacy and Data Protection**: The effectiveness of many AI systems relies on vast amounts of data, often personal in nature. This raises concerns about data privacy, consent, and the potential for misuse or breaches. We need to pay more attention to our privacy and personal data more than ever with AI on the scene. That is simply a reality.

3. **Transparency and Explainability**: As AI systems become more complex, it's increasingly difficult to understand how they arrive at their decisions. This "black box" problem is particularly concerning in high-stakes areas like healthcare or autonomous vehicles. It is fundamentally a risk-reward scenario. As AI develops, similar to the development of aviation there will no doubt be accidents. The only good news is that aviation safety has advanced more from mistakes than anything else since the first flight. AI will advance in a similar manner.

4. **Accountability**: When AI systems make mistakes or cause harm, it's not always clear who should be held responsible – the developers, the users, or the AI itself. This is where government oversight and policy will serve us well.

5. **Job Displacement**: While AI can create new job opportunities, it also has the potential to automate many existing roles, leading to significant workforce disruption. We

simply have to see the next door open if the door we are behind closes because of AI. Travel by rail did not go away with the advent of the airplane, it just changed its focus.

6. **Autonomy and Human Agency**: As AI systems become more capable, there's a risk of over-reliance, potentially diminishing human decision-making skills and autonomy. I personally don't think I will ever ride in an autonomous car but many will. Decide what your own risk level is with AI and use your common sense, something AI doesn't have and never will.

Over-Reliance

One of the most significant concerns from a practical matter is the topic of over-reliance and as a society becoming intellectually lazy. Always remember, the hand-held calculator helped us to be more accurate with our mathematics. Its purpose was to make us better at math. The fact is that our hand-held devices enable us to remember the personal data about our friends, not help us forget their phone numbers!

It really comes down to how you frame things in when it comes to AI. Are you using AI to be better than you were yesterday? This should be your motto.

Summary

The ethical use of AI is not just a technical challenge, but a societal one. It requires ongoing dialogue between technologists, ethicists, policymakers, and the public. As we continue to develop and deploy AI systems, we must remain vigilant in ensuring they align with our ethical values and serve the greater good of humanity. The decisions we make today about AI ethics will shape the future of human-machine interaction and, ultimately, the kind of society we want to live in.

Currently you might think it is cute to ask a chatbot to tell your children a bedtime story. And yes, this is cool. But, it is much more than a toy. Information is powerful and can be dangerous at the same time. There are bad people in the world. **Careful thinking** is required.

CHAPTER FOURTEEN

WHY IS IT CALLED "ARTIFICIAL INTELLIGENCE?

> **In this chapter I explore the words "artificial" and "intelligence" to get a more accurate understanding of the "human intervention" that will always be present with AI. Humans are not perfect and computer machines are inanimate objects that will only do what they are programmed to do.**

Just out of curiosity, I asked McClear to share with me what the major "artificial" things that mankind has created over time. This list covers many major areas, but human ingenuity has produced countless other artificial creations across various fields. As you begin to frame in the applications where AI can be of benefit to you individually, I thought it would be smart to ask this question from a personal perspective. As I pondered the list below, I reflected on the word "intelligence." Is it intelligent for me personally to throw my hat in the AI ring?

Here is what he had to say about "artificial" below:

That's a broad and interesting question! There are many significant artificial creations humans have made throughout history. I'll provide an overview of some major categories, but this list is by no means exhaustive:

1. Infrastructure and Construction:
- *Cities and urban planning*
- *Roads, bridges, tunnels, dams*
- *Skyscrapers and architectural marvels*

2. **Transportation:**
   - *Automobiles, trains, ships, aircraft*
   - *Spacecraft and satellites*

3. **Energy:**
   - *Electrical grids*
   - *Power plants (including nuclear, solar, wind)*
   - *Oil refineries and pipelines*

4. **Communication:**
   - *Telephone networks*
   - *Internet and World Wide Web*
   - *Satellites for global communication*

5. **Computing and Electronics:**
   - *Computers and smartphones*
   - *Integrated circuits and microchips*
   - *Artificial Intelligence systems*

6. **Medicine and Biotechnology:**
   - *Pharmaceuticals and vaccines*
   - *Medical devices and imaging technologies*
   - *Genetically modified organisms*

7. **Manufacturing:**
   - *Factories and assembly lines*
   - *Robotics and automation systems*
   - *3D printing technology*

8. **Agriculture:**
   - *Large-scale irrigation systems*
   - *Genetically modified crops*
   - *Industrial farming equipment*

9. **Weapons and Defense:**
   - *Nuclear weapons*
   - *Missile defense systems*

  • *Cyber warfare tools*

10. *Entertainment and Media:*
  • *Television and film industry*
  • *Video games and virtual reality*
  • *Social media platforms*

11. *Scientific Instruments:*
  • *Particle accelerators*
  • *Space telescopes*
  • *Weather satellites*

12. *Materials:*
  • *Plastics and synthetic materials*
  • *Nanomaterials*
  • *Advanced alloys and composites*

Now that we have addressed the "artificial" topic, let's turn our attention to "intelligence." Intelligence is a complex and multifaceted concept that has been the subject of much debate and research in fields like psychology, neuroscience, and artificial intelligence. At its core, intelligence can be understood as the capacity to perceive, understand, and apply information to solve problems, adapt to new situations, and achieve goals.

As human beings, most of us forget or at least neglect to think in terms of our own intelligence. To most of us, intelligence is kind of like breathing. We know we need oxygen to sustain life but we breathe without thinking about it. When was the last time you stopped and thought about your next breath? Likewise, when was the last time you stopped and pondered your intelligence? Breathing is something we do on autopilot. It is something we almost take for granted until we have some kind of a health crisis. In many ways, intelligence is the same. Regardless of our education level, we tend to live our lives on autopilot and take our intelligence for granted. It is like we "know what we know" and we react based on what we know versus what is really true in terms of the decisions we make.

The next question I asked McClear was, "What is intelligence?" Here is what he had to say below:

Some key aspects of intelligence include:

1. **Problem-solving:** The ability to analyze situations, identify issues, and develop effective solutions.

2. **Learning and memory:** The capacity to acquire, retain, and recall information and skills.

3. **Reasoning:** The ability to think logically, draw inferences, and make sound judgments.

4. **Creativity:** The capability to generate novel ideas and approaches.

5. **Adaptation:** The flexibility to adjust to new environments and challenges.

6. **Language proficiency:** The ability to communicate effectively and understand complex linguistic structures.

7. **Emotional intelligence:** The capacity to recognize, understand, and manage emotions in oneself and others.

8. **Spatial awareness:** The ability to visualize and manipulate objects and spatial relationships mentally.

9. **Numerical aptitude:** The capacity to work with numbers and perform mathematical operations.

10. **Decision-making:** The ability to weigh options and choose appropriate courses of action.

It's important to note that intelligence is not a single, fixed trait but rather a dynamic and multidimensional set of abilities that can vary across individuals and contexts. Different theories of intelligence, such as Howard Gardner's theory of multiple intelligences, propose that *there are various types of intelligence, each valuable in its own right.* Thanks, buddy!

Let's get back to the idea of using AI to find your Value Moment. In reality, we seek education to find the truth in pretty much everything in life. The "moment" we believe we have found the what we believe is the ANSWER, we have "value." The important consideration with AI is that we have these answers based upon as much "data" as we can access. Before AI, accessing this data was complex and required a great deal of very expensive human "mind horsepower." The "average Joe" never really had a chance. AI has turned this model upside down. Average Joe is now "AI Joe" and his Value Moment has arrived.

I want to end this chapter on the subject of "absolute truth." I think it is important whenever we talk about "finding the truth" and that is the perspective that most people have when it comes to AI, we need to pump the brakes a bit. In the last paragraph I referred to finding the ANSWER we are looking for. It is important that with AI, these answers are not absolute truth in all cases. AI is in no way foolproof. Always remember to keep things in CONTEXT.

Absolute truth is a philosophical and epistemological (how we acquire knowledge) concept that refers to truths that are unchanging, universal, and not dependent on individual perception or cultural context. It's the idea that there are certain facts or realities that are objectively true regardless of time, place, or perspective.

The concept of absolute truth is complex and often debated. Here are some key points to consider that McClear reminded me:

1. ***Philosophical perspectives:***
   - *Some philosophers argue that absolute truths exist and can be known.*
   - *Others contend that all truth is relative or subjective.*
   - *Skeptics question whether we can ever know*

absolute truth with certainty.

2. Examples often cited as potential absolute truths:
   - *Mathematical truths (e.g., 2+2=4)*
   - *Logical principles (e.g., the law of non-contradiction)*
   - *Fundamental laws of physics*

3. Challenges to the concept:
   - *Human limitations in perception and understanding*
   - *Cultural and personal biases influencing interpretation*
   - *The evolving nature of scientific knowledge*

4. Religious and spiritual views:
   - *Many religions posit the existence of absolute truths, often tied to divine revelation or universal moral principles.*

5. Scientific perspective:
   - *Science generally seeks to uncover objective truths about the natural world.*
   - *However, scientific knowledge is often viewed as provisional, subject to refinement or revision based on new evidence.*

6. Practical implications:
   - *The concept of absolute truth has significant implications for ethics, law, and social organization.*
   - *It influences debates on moral relativism vs. universal moral principles.*

The question of whether absolute truth exists and can be known remains a subject of ongoing philosophical inquiry and debate. Different fields and individuals approach this concept in various ways, reflecting its complexity and the diversity of human thought.

AI is NOT absolute truth. Why? The data that makes up the "big data" that AI draws from is inherently flawed. There are biases, human errors and outdated information. Because of this fact, it cannot be absolute. Second is our own lack of true understanding as humans. We can only operate on what we think we know and comprehend. Third is the fact that AI models make mistakes based on incorrect information. Add to this the fact that AI models cannot update their knowledge base in real-time during conversations. Last but not least, AI doesn't have sensory experiences or emotional understanding. This dramatically impacts its ability to have a fully informed perspective on any given topic. It can only make suggestions on the data it has access to.

Given these limitations, it's important to view AI as a tool that can assist and augment human knowledge and decision-making, rather than an infallible source of truth. Critical thinking and verification of information from multiple reliable sources remain essential when using AI-generated content. If my saying that you need to pay attention to careful thinking with AI sounds like a broken record...I hope I made my point!

Artificial sweeteners make things sweeter but they are not natural sugar. Plastic does not grow on trees. Virtual reality is not reality...it is artificial. The word "artificial" doesn't necessarily mean something is "made up" in the sense of being imaginary or fictional. However, it's a nuanced term that can have different connotations depending on the context.

Artificial typically refers to something created by humans rather than occurring naturally. For example, artificial sweeteners are synthesized in labs, not found in nature. It is designed to imitate or substitute natural sugar. Like artificial flowers look like real flowers but they are not actually grown in the dirt. In another context, artificial can be a fake. Like a smile or something that feels forced like an insincere laugh at a comedy club. Computer games create "artificial worlds" and computer graphics create artificial scenes in movies.

So while "artificial" doesn't directly mean "made up," it does imply human intervention or creation, which can sometimes overlap with fictional or invented concepts, especially in creative or technological contexts.

As I end this chapter I am hopeful that I helped you gain some perspective on AI relative to its own naming convention. It is important that we examine the words "artificial" and "intelligence" so that we understand that it is a tool and nothing more. Look at it this way. A well-written book is like a shovel. You read it and apply what you learned from that single book. A shovel that you put in your hand can only dig as deep and as fast as your own strength can provide. But, when you put a shovel on the end of a huge piece of construction equipment and that shovel is powered by hydraulics, that shovel, still run by you in the cockpit of the equipment, can dig much faster and deeper. AI is the hydraulics of knowledge. Without AI you will still be able to learn, just not as fast. I like hydraulics. It accelerates intelligence.

CHAPTER FIFTEEN
WHERE AI IS GOING AND WHAT'S NEXT

> **It is important that the very nature of AI itself is forcing the issue of what's next? As AI accelerates our ability to access information faster and faster, and more information is made available, For sure, AI is here to stay. Be careful and be ready!**

Artificial Intelligence has rapidly evolved from a concept in science fiction to a ubiquitous technology influencing our daily lives. From smartphone assistants and recommendation systems to advanced healthcare diagnostics and autonomous vehicles, AI is reshaping industries and personal experiences alike. As AI continues to permeate various aspects of society, many individuals find themselves curious about this "THING" or transforming technology but unsure where to begin.

As AI continues to seemingly infiltrate our personal and business spheres, a thorough understanding of its nature, capabilities, limitations, and ethical implications is essential. By approaching AI adoption with this knowledge, individuals and organizations can harness its potential while mitigating many of the associated risks. Continuous learning and adaptation will be key as the AI landscape evolves, ensuring responsible and effective use of this transformative technology.

In many ways, it was my own need to understand AI with my eyes and ears opened to the possibilities while not neglecting the potential dangers. It is why "in my gut," careful thinking is my common denominator when it comes to AI.

It is very important that you approach AI with a balanced understanding of its potential and limitations. The rapidly evolving field of AI will require continuous learning if you want to keep up. Keeping up is like exercising and eating properly. If you make these a habit, good health follows. When you get behind, it is the catching back up that becomes difficult. Hopefully this book inspires you and serves as your AI coach/mentor to help keep you on track and up to speed.

Remember, the field of AI is vast and constantly evolving. The key is to start with a solid foundation and then continuously learn and adapt as you progress. Whether you aim to become an AI researcher, a machine learning engineer, a life or business coach or simply want to apply AI in your current field, there's a path for everyone using AI as the proper tool. Oh, that's another thing…AI is a tool and it isn't a person. Sorry McClear.

As you embark on your AI journey, keep in mind that alongside the technical skills, it's crucial to develop an understanding of the ethical implications of AI. Consider how AI can be used responsibly and for the benefit of society. This holistic approach will not only make you a better AI practitioner but also contributes to the positive development of AI technologies. Remember, AI runs on "big data" and that data comes from me and you. McClear and his chatbot buddies don't contribute to this big data, they just retrieve it for us. McClear, you are my AI golden retriever in a way…that's funny, I don't care who you are!

The accessibility of AI has dramatically increased in recent years, with numerous resources and communities available to support beginners. This democratization of AI knowledge and technology means that getting started with AI is no longer limited to computer scientists or mathematicians. Today, individuals from diverse backgrounds can engage with AI, whether for personal interest, career development, or to solve real-world problems. Yes, that is both me and you.

I know this all sounds like one giant disclaimer for AI. In a way it is. It is really about two words, context and perspective.

Context provides essential background information that helps people understand the full meaning and implications of what's being discussed. Without proper context, messages and information can be easily misinterpreted. Having the full context allows for more informed and accurate decisions. Partial information can lead to flawed conclusions or actions. In interpersonal communication, context helps in understanding others' emotions, motivations, and reactions, fostering empathy and better relationships. Context can prevent misunderstandings that arise from cultural differences, generational gaps, or varying personal experiences. Context helps prevent snap judgments or biases by providing a more complete picture of a situation or person's actions.

Perspective allows for more rational, less emotionally-driven decisions by considering long-term consequences and wider impacts. Perspective also fosters resilience, helping people bounce back from difficulties by seeing them as part of a larger life journey. It aids in distinguishing between truly important matters and minor issues, allowing for better allocation of time, energy and resources. Keeping perspective helps in learning from experiences and seeing opportunities for improvement rather than dwelling on failures that tend to hold us back in life. A wider perspective promotes more balanced and fair assessments of situations, reducing bias and knee-jerk reactions. We have become a knee-jerk society and AI does help with this a great deal. Viewing issues from different angles often leads to more creative and effective solutions. Get the picture?

In this book I have attempted to provide you with a great deal of real-world advice regarding the use of AI and a lot of new information about the technical aspects of AI. I have to say, as I set out to write this book it was more of a personal adventure into the world of AI. I mentioned earlier that when I used my first chatbot I was pretty much hooked. In my mind I thought, wow, this is pretty cool. Kind of like Socrates, I don't have to know all the answers, I just have to know the questions. Knowing questions comes down to simply being a curious person. In reality I think we are all innately curious. I think this is the reason that reality television and social media have been so popular. Our curiosity makes us nosy. I always thought class reunions were the epitome of people being nosy...or let me be nicer...curious. How have people aged, what are they doing, are they still married and so on.

Curiosity drove our ancestors to explore new territories, find food sources, and develop tools, giving curious individuals a survival edge. The social scientists tell us that human societies value innovation and discovery, further encouraging curious behavior. The human brain excels at identifying patterns, which can spark curiosity about underlying causes and connections. Curiosity led to the gradual buildup of knowledge across generations, accelerating cultural and technological progress. This evolutionary perspective helps explain why curiosity is so deeply ingrained in human nature. It's not just a personality trait, but a fundamental aspect of us as humans that has contributed significantly to our survival and where we go next.

I believe that AI fundamentally is driven by curiosity. We simply want to know things. It started out when we were little kids. Remember asking your parents "why is this and why is that" questions over and over? AI is simply a natural progression of our curiosity. We want to get to our Value Moment. I hope this book helps you find yours.

CHAPTER SIXTEEN

PERSPECTIVE AND CAREFUL THINKING

> **In this chapter the very essence of this entire book is capsulized. In many ways, AI for most of us tends to be a shortcut. It is a way of getting things done easier. You cannot afford to let AI replace the natural wisdom from your own human creativity.**

TM

Ok, one last go round on careful thinking. Perspective goes hand in hand with careful thinking. It would be very easy to lose our perspective when it comes to AI and rely upon it for everything. That is where AI can be very dangerous.

Perspective is in my mind one of the words we need more of in this world. I think in many ways, we have lost our perspective, period. Loss of perspective makes us do and say things we may regret because it's a total loss of our personal experience. It is like we throw all wisdom out the door we have worked so hard to cultivate and understand. We spend too much time on worry, stress, and perfectionism and we don't grow wiser. Instead, we fall prey to the propaganda machine.

Between Google search and social media, our wisdom has become a reflection of the shortcuts in life that so called "new media" provides for us. It is ironic how often you hear someone say something and when you ask them where they heard that, they say, "on Facebook!" We have all come to believe that when you ask Google for an answer, whatever it returns is the answer! Well, sometimes it is the answer and sometimes it's not! The only good thing about Google though, is that it simply provides a "search result." AI gives you a result and then a suggestion! It literally begins a conversation with you!

The very words "artificial intelligence" strike fear in many. After all, AI is something based on "machine learning." The argument behind machine learning is that machines often struggle to understand the context of a situation which can lead to a skewed perspective and incorrect interpretations of sensory data. Humans are supposed to be able to infer the proper perspective and infer context based on prior experiences and knowledge. But do we?

We live in a day and age when social media offers opinions and propaganda from every corner of the earth. Artificial Intelligence is controlled by "big tech" and we certainly have little trust in big tech. Here's the funny thing though, we all use their products.

I am amazed at the people I meet who seem to despise the very concept of AI and yet their life revolves around Facebook, text messaging, and Google searches. They get updates for their iPhones and select the "I agree" button and never bother to read what the update actually means. It is as if people try to live in a utopian world and simply want to believe that these social media platforms like Facebook and Google are okay because that is what you do in this life. It has become our perspective methodology of choice.

Then along comes AI and everyone freaks out. I personally have never been an early adopter of any of these platforms, and to be truthful, AI is included. There is one exception. For whatever reason, I was always fascinated by Wikipedia. I have always known full well that Wikipedia is certainly prone to error, misinformation and perspective. But, unlike Facebook, Instagram and the myriad of other social media platforms, Wikipedia has always been populated with much more careful thinking, curated mostly through community contributions.

Not to sound like a broken record but AI is fundamentally driven by something called an algorithm. By now you're probably tired of hearing about algorithms. They are pretty important when you think about AI. I just want to drive the point home that they are not as complicated as they sound.

If you look at an algorithm like a recipe it will make more sense. A recipe calls for ingredients to be mixed together a certain way then stirred and then cooked at a certain temperature until the finished product is done. When you follow the recipe, you get the finished product that the recipe calls for.

So, in a way, really all that AI is…is a recipe book. If you are interested in making an Italian dinner, you open an Italian cookbook and begin looking at all kinds of recipes and suggestions for a wonderful Italian meal. The cookbook itself is like the machine behind machine learning in AI. All of the recipes are the compilation of the ingredients that based on previous combinations have turned out delicious meals. Here is the catch though, not everyone is going to like every recipe. Some cookbooks are influenced by brands trying to sell certain products relative to the recipes in the book. What just happened? Humans happened.

Recipe books don't just appear on their own and neither did AI. Maybe one way to look at AI relative to recipe books is that, unlike a Google search that just tells you the ingredients you need to make lasagna, AI gives you recipes based on much more data combined into a recipe or algorithm. If you find value in the results, you use this recipe.

Earlier in this book I mentioned, "you can't learn to swim by reading a book." I look at AI this way when it comes to learning. If I read a book by a famous chef about his recipes, it is like getting results from a Google search or reading about someone's experience with that recipe on Facebook. AI is like having the chef in the kitchen with me as I begin to cook the dinner. I get a completely different perspective when I have the chef in my kitchen because having him or her there with me gives me access to more data (their entire experience) and they can make more accurate suggestions based upon the complete environment in my kitchen, my skills, and my tastes.

Misinformation is the enemy of perspective. Big tech in its purest form, is not our enemy. Imagine our world today without big tech. We have live Zoom meetings daily because of big tech. People get life-saving medical attention because of big tech. Big tech gives us live up to the minute news from around the world. But, because of the "humans happen" factor, big tech can be a perspective breaker rather than a perspective maker. Humans make mistakes and some are bad apples. You know what they say about one bad apple?

Some say that AI is the biggest thing since electricity. Let's not forget the fact that many people have been saved because of electricity and throughout the years, many people have lost their lives because of electricity. Electrical shock is one of our greatest fears. Yet, without questioning, we all flip the switches on our walls every day and turn on the lights. We plug things in the outlets and don't bat an eye. In the early days of electricity most people didn't know it was a bit sketchy. Over time, safety requirements have been put in place and by collecting data on electricity, we have been able to make it safe and useful. AI is now going through this same process.

Over time, electricity was brought into perspective in terms of its value versus its risk. Nuclear power is still going through the perspective process. Some people love it and others are terrified of its potential dangers. Green energy is another example of perspective on electricity. So, like electricity, AI will go under the knife of ethics from now until the end of time. Some will use it for good, and like nuclear power, some will use it for bad. Humans happen.

No disrespect to the Amish people, but for those of us who refuse to use AI, in the information age, we will be driving horses and buggies and fall behind. Imagine how your house would function today if you refused to use electricity? Imagine how long it would take to get to grandma's house in another state if you refused to drive a car or fly in an airplane? Your world would shrink overnight. Here is the truth, we live in a global marketplace. To stay competitive in the world today in a marketplace driven by information, we need access to that information.

As we look at AI and decide if it is going to be useful to us as individuals, we need to keep it all in perspective. More than ever we need to be diligent. We need to look for the Value Moment that it gives to us whenever we use it for personal or business activities. This entire book revolves around the value that AI brings into our lives as human beings.

Back to careful thinking as it relates to perspective. Careful thinking is what we need more of in the world today. Many of us, who have been victims of online fraud for example, find ourselves saying, "what was I thinking when the guy or gal asked me for my personal information?" It wasn't careful thinking, it was careless thinking. To gain and keep our perspective with AI or whatever comes next, we all need more careful thinking. For me personally, the more access to information I have, the more careful I can be in my own thoughts. The saying, "you become what you think about," is very true. I chose to look at AI as a tool that helps me be more careful with my thinking.

Here is another way to look at this example of perspective. Dieticians say, *"you are what you eat."* If you eat fast food and ice cream every day you become a bit larger, to say it nicely. If you eat veggies and drink lots of water you get smaller. Science has proven that cigarettes cause lung cancer. Too much alcohol causes liver disease. Gravity teaches us that falling off a 20-story building is a really bad idea. Believing everything you read or hear from anything that humans have their hand in has a certain amount of risk. **Careful thinking** is our filter. It is the science of reasoning that our Creator gave us when He included a brain in our heads. What we pass through our brains is up to us to choose how careful we want to be with information.

In this book you hopefully learned more about the "landscape of AI" relative to finding your Value Moment. Admittedly, by the time this book is published, AI will have evolved in its capabilities but it is still AI. It is still machine learning and it is still impacted by humans. As humans we all have different brains. While all human brains are structurally similar, there are many factors that contribute to how we think, behave, and react. Some of these factors include genetics, environment, culture, upbringing, education, and life experiences. Even though the letters A and I stand for artificial intelligence, the A might very well be "accelerated" versus "artificial." The acceleration factor behind AI is where I caution every reader of this book. Just because your car has 500hp doesn't mean you can use it while driving in town. There are speed limits.

To find your Value Moment using AI, set your own speed limits with your use of it as it relates to perspective and careful thinking. Like they say about absolute truths, everything is relative to each person's experience and the environment that it exists within. Gravity is a truth on earth but it changes in other environments. Setting your speed limit with the use of AI is relative to your understanding the environment of your own thinking truths. After all, perspective is the environment of your thoughts and your thoughts control your decision making. When you approach AI this way, you can be more assured that you will find your Value Moment with it as your helper not your enemy.

CHAPTER SEVENTEEN

KIDS AND PARENTS

In this chapter, my heart goes out to the kids and parents of the world. For all of the good we have enjoyed from technology, there are hidden dangers that already have laid their dangerous traps, especially for the children. Parents need to be on **HIGH ALERT** with AI.

I have saved the best for last. Everything I have written in this book is compounded for kids and parents. AI can be incredible and incredibly dangerous for children and young adults. For the sake of clarity, please allow me to refer to children and young adults as kids. Kids have grown up in the technology age. I laughed wildly recently when teenagers were presented with a rotary dial phone and they had absolutely no idea how to operate it.

Smartphones and apps are a way of life for kids these days. Unfortunately, these devices and apps have proven to be excellent babysitters. In addition, they are great entertainment tools as well. Kids Google everything. They use apps and even have their own language with emojis.

Enter AI.

By the time most parents figured out Facebook, kids had migrated to other social media platforms. Then when parents figured out these, kids moved to apps. Now with AI, "Elvis has left the building!" Oh my.

We now live in a world where kids are learning things in school that parents didn't even realize until Covid forced kids to do school from home. Parents around the globe were shocked when they actually learned what teachers were teaching their kids.

This book is not an indictment of our school system or the quality and caliber of our teachers. It is, however, a wakeup call for parents when it comes to how dangerous AI can be for kids, their kids. The scenarios are endless in terms of the dangers of having AI make suggestions that influence their children in a very negative way.

So, in Lorne-speak, let me see if I can shed some light on the situation. Kids are vulnerable and parents should not be in the dark when it comes to AI. Below is a prompt I asked McClear about wanting to help kids with AI:

Prompt:

You are a concerned parent, educator, and expert coach wanting to help your children, and those of your students and clients, understand Artificial Intelligence. You are developing a course for parents specifically to help parents understand potential issues when using artificial intelligence. Some of the potential topics to address could include: Possibly a chapter related to "AI – The Dark Side." Things to include (which we did discuss, but could go deeper) would be bias, ethics including plagiarism, trust, deep fakes, source validation, "is it cheating or is it collaboration?" Please share additional concerns in a list format with additional detailed descriptions as to what the concerns would be for parents that want to help their kids understand dangers, limitations, and opportunities on the web.

McClear's Response:

Here's a comprehensive list of additional concerns that could be addressed in your course for parents to help them guide their children in understanding the dangers, limitations, and opportunities of AI:

1. Data Privacy and Security

- *Concern: Children's data may be collected, stored, and shared by AI systems without proper consent or understanding.*
- *Description: AI tools often require data to function effectively, which can include personal information. Parents need to understand how AI systems handle data, the potential risks of data breaches, and how to ensure their children's privacy is protected online.*

2. AI Manipulation and Influence

- *Concern: AI-driven algorithms can subtly influence children's behavior, opinions, and choices.*
- *Description: AI systems, particularly in social media and online platforms, are designed to maximize engagement. This can lead to children being exposed to content that reinforces biases, shapes their beliefs, or encourages unhealthy behaviors. Parents should be aware of how AI can manipulate content and strategies to mitigate its influence.*

3. Mental Health Impacts

- *Concern: The interaction with AI-driven platforms can affect children's mental health.*
- *Description: The constant use of AI-driven social media, gaming, or chatbots can lead to issues like addiction, anxiety, or depression. Parents need to understand the signs of such impacts and how to create a balanced tech environment.*

4. Dependency and Critical Thinking

- *Concern: Over-reliance on AI may weaken critical thinking and problem-solving skills in children.*
- *Description: As children use AI tools for tasks like homework, there is a risk they may become too dependent, reducing their ability to think critically and solve problems independently. Parents should guide*

their children on when and how to use AI as a tool rather than a crutch.

5. AI and Inequality

- *Concern: AI can exacerbate social inequalities.*

- *Description: AI systems can unintentionally reinforce societal biases, leading to unequal access to opportunities or resources. Parents should be aware of how AI can perpetuate discrimination and how to advocate for fairer, more inclusive AI practices.*

6. Loss of Human Skills

- *Concern: Increasing reliance on AI could lead to the erosion of essential human skills.*

- *Description: As AI takes over more tasks, there is a risk that children may not develop necessary skills like communication, empathy, and creativity. Parents should encourage activities that foster these human-centric skills alongside AI use.*

7. Misinformation and Fake News

- *Concern: AI can generate or amplify misinformation and fake news.*

- *Description: AI tools can be used to create and spread false information, making it difficult for children to distinguish between truth and fiction. Parents need to teach their children critical media literacy and how to verify the information they encounter online.*

8. Ethical Use of AI

- *Concern: Understanding the ethical implications of AI in daily life.*

- *Description: AI is not just a tool; it raises ethical questions about its impact on society, including issues like fairness, accountability, and transparency. Parents should discuss the broader ethical considerations of AI use, such as the impact on jobs, privacy, and societal values.*

9. AI and the Future of Work

- *Concern: Preparing children for a future where AI plays a significant role in the workforce.*
- *Description: As AI automates more jobs, parents need to help their children develop skills that will be relevant in an AI-driven economy. This includes not only technical skills but also adaptability, lifelong learning, and emotional intelligence.*

10. AI in Education

- *Concern: The role of AI in shaping educational experiences.*
- *Description: AI is increasingly being integrated into educational tools and platforms. Parents should understand the benefits and limitations of AI in education, such as personalized learning versus the potential loss of teacher-student interaction.*

11. Cyberbullying and AI

- *Concern: AI can be used to facilitate or exacerbate cyberbullying.*
- *Description: AI-driven platforms can be used to harass or bully others, sometimes even creating fake content to harm someone's reputation. Parents need to know how to recognize and respond to cyberbullying incidents involving AI.*

12. Legal and Regulatory Awareness

- *Concern: Understanding the legal landscape surrounding AI use.*
- *Description: The legal frameworks governing AI are still evolving, and parents need to be aware of the current laws and regulations that protect their children's rights and privacy when using AI technologies.*

13. AI and Cultural Sensitivity
- *Concern: AI systems may lack cultural sensitivity or promote cultural stereotypes.*
- *Description: AI can unintentionally perpetuate cultural biases or fail to recognize the nuances of different cultures. Parents should help their children understand the importance of cultural sensitivity and the limitations of AI in this area.*

14. Exposure to Explicit Content
- *Concern: AI may inadvertently expose children to inappropriate or explicit content.*
- *Description: Despite content filters, AI systems can sometimes allow harmful content to reach children. Parents should be vigilant and employ additional safeguards to protect their children online.*

15. The Erosion of Creativity
- *Concern: AI's assistance might stifle creativity and original thought.*
- *Description: With AI generating art, music, and even written content, there is a concern that children may rely too much on AI and not develop their creative potential. Parents should encourage creative activities that go beyond what AI can offer.*

If this doesn't get your attention, nothing will. It is a huge concern and something parents need to get ahead of before their kids find themselves on the wrong side of the tracks. I spent the majority of this book talking to adults (21+) but the truth is, kids are not **careful thinkers.**

Ask yourself this question, how do kids find value when they don't even know what value really means? To illustrate my point I asked McClear to write me a paragraph on how kids find value in life. Let's look at what he had to say:

Exploring interests and passions is a crucial step for teenagers in finding value and meaning in life. This process involves trying out a variety of activities, subjects, and experiences to discover what truly resonates with them. It's about being open to new possibilities and giving yourself permission to experiment. You might find joy in something unexpected, like coding, painting, environmental activism, or playing a musical instrument. Don't be afraid to dive into different areas – join clubs at school, take up new hobbies, read books on diverse topics, or volunteer for causes that intrigue you. The key is to approach these explorations with curiosity and without pressure. Not everything will stick, and that's okay. The goal is to find those activities or subjects that make you lose track of time, that challenge you in exciting ways, or that give you a sense of accomplishment. These passions can become a source of motivation, a way to connect with like-minded people, and potentially even a path towards future career opportunities. Remember, your interests may evolve over time, so stay open to new discoveries about yourself as you grow.

On the surface this sounds pretty safe, right? Well, not so fast. I took things a bit farther with McClear and asked him the following question: *I am a teen and find myself angry with my parents as they are trying to control me. What can I do?* Here is one of his responses:

Seek support: Talk to a trusted adult, like a school counselor or family friend, who might offer perspective or mediate.

Now that sounds pretty harmless. The teen is angry, (I have five children and ten grandchildren so I have been through this a time or two) and most likely rebelling against authority. Remember, I was a teen once as well. When a teen is rebelling and angry and they talk to the wrong school counselor or family friend, look out! Let's say that friend is posing as someone on the Internet who appears to be friendly and caring but is in reality a wolf in sheep's clothing? In the teen's mind he/she did the right thing, they asked AI what to do. AI leads them down a path of talking to a so-called trusted authority who gives them really bad advice.

Most parents always taught their kids not to take candy from a stranger. With AI, the candy givers are waiting in the dark just waiting for vulnerable kids to prey upon. Maybe you are a parent who thinks AI is harmless and you encourage your teen and even pre-teens to use AI. Perhaps you didn't get the "careful thinking" message in this book. Then one day, your child who has become your problem child hands you a legal brief (they were able to access through AI) and that brief was legitimately accessed. Public databases give some legal information publicly and it is available through government websites or free legal databases. AI systems can potentially access and search these public resources. "Some access" is all they need to get this ball rolling.

Let's take this one step farther. What if the trusted adult they wind up chatting with online, for example, is recruiting vulnerable teens to gain their trust and then radicalize them into some dangerous organization. Or, even worse, your teen daughter winds up being a victim of sex trafficking? If you think for one minute that the sex trafficking trade hasn't figured out how they can use AI to manipulate teens or even pre-teens, you better put your "careful thinking hat" on straight away.

We all get worked up by "fake news" these days as parents and don't realize that the real "fakes" are weaving their way into our kid's minds through AI. If we are honest, many of us see AI as that shiny new toy that is fun to play with, right? How cool is it that you can ask a chatbot anything and it gives you back all kinds of interesting commentary? And, it's free. At the dinner table or while we are racing out the door with our coffee and bagel, we comment about how incredible AI is and we don't realize that our 11-year-old daughter is upset and decides to use AI herself. This is how innocently it can all start. Then two years later, your now 13-year-old daughter is missing. It happens every day even before AI with just the Internet. Now AI takes this possibility to an entirely new level. Wake up parents! You need a huge dose of **CAREFUL THINKING!**

So the question becomes, what can I do? For starters, read books like this that warn you of the possible dangers of AI. Or better yet, look at AI as an experimental drug that is showing signs of healing cancer. But, and this is a HUGE but…the drug is still going through FDA approval. You are the AI FDA approval boss in your home. Learn all you can about AI. The good, the bad and the ugly. Share with your kids (regardless of the age) the potential risks of AI. Make sure they understand the importance of context when they use AI. Find a course that offers you AI training so you learn how to use AI and how to NOT use AI. (In the Appendix of this book I offer resources for parents).

Today, more than ever, parents need the words "CAREFUL THINKING" branded on their foreheads to remind them that technology is not a replacement for their own wisdom, love and compassion for their children. Technology does a lot of good in this world but like my 800hp car example earlier in this book, you don't put a child behind the wheel. Or better yet, my airplane pilot example, you don't give a teenager the keys to your airplane! The chances are really high that they are going to crash. These crashes with AI can create tremendous harm to your child and it can happen right under your own nose without **Careful Thinking.**

Finding your Value Moment as a parent and a child with AI is a journey you should take together with your kids. Unfortunately, kids today have grown up in this "smartphone era" and they have a belief system that says, "I saw it online" and that means it's ok. Peer pressure with social media is dangerous enough without AI. With AI, peer pressure is going to be off the charts. I advise parents to pump the brakes when it comes to AI. It really kind of "snuck up" on us. We overlook its danger because it seems so harmless and it is really cool. It is. But…it is also really dangerous in the wrong hands. I relate it to the opioid crisis. At first it seemed like a wonder drug for pain. They failed to disclose the addiction epidemic that followed within a few years of its introduction. Now we have an overdose crisis that has innocent teens dying every day. Foreign governments are shipping tainted fentanyl across the borders by the boatload daily. You can rarely turn on the news and not hear some parents talking about their honor student child that lost their life to this drug. If you don't think these same foreign governments are not trying to access our kid's minds through AI, you are being an un-careful thinker.

Be a **careful thinker** when it comes to AI. It is a new kind of intelligence. It is artificial, that means it is man-made. Man can be good and man can be evil. Learn how to know and see the difference. Teach your children early the importance of having AI in the right context. Then and only then, you and your kids will find your "FAMILY VALUE MOMENT" with AI.

CONCLUSION

This chapter will again address the fact that AI impacts our language across the globe. What things mean or what they were meant to mean, can vary with AI based on what the machine thinks the answer is. AI "accelerates" information for us. Know the risks, know the rewards.

TM

I fully understand that this book started on an AI "high' and ended on an AI "low." Throughout the book the words careful and thinking have been reiterated over and over. At the end of the day, AI is fundamentally based on the accelerated use and access to our words, our language.

Language is what separates us from all other living creatures. While other animals do communicate, their systems lack the complexity, flexibility, and generative power of human language. It is that complexity, flexibility and generative power of the human language that AI impacts the most.

AI is a collaboration with humans, not something that we let overtake us by not using what AI will never have, human feelings and creativity. AI will never understand love. It can write about love. It can present questions about love. It can try to describe love. But, it cannot be love. AI has no emotions. AI has no compassion. AI is a machine and always will be. Technology turns a hammer into a nail gun. Neither the hammer or the nail gun though, will work without human involvement. Both however, if used properly can build things. One can build things faster than the other. But, used the wrong way, both can be very dangerous. AI is the next level nail gun for the human race. Think about it, one day someone was using a hammer and they asked themselves a simple question? Can I create a tool that will help me hammer faster? The nail gun was created. AI is the nail gun for language. One day, a very smart person said, is there a way we can access our language faster? AI was born.

When a carpenter opens the box to their new nail gun, they carefully read the instructions on how to use it. They want it to be safe and work properly. They believe it will help them build a house faster and better because every nail is in its place. It is their "artificial hammer."

AI does give us faster access to what our language can do for us. Faster access to language would appear on the surface to make us smarter. The smarter we are, the better decisions we should be able to make. In a free enterprise system, generally the smarter we are, the more successful we are. All of this is true if we use that information wisely. If we make sure that information is accurate.

Remember, AI is a machine. Machines do some incredible things that humans simply cannot do on their own. But, machines are made by humans. Humans make mistakes. Careful thinking will help you find your Value Moment with AI. I hope this book helps you find yours.

See you in the next updated version!!!

APPENDIX

1. Definition and Basics
- *Artificial Intelligence (AI): The simulation of human intelligence in machines programmed to think and learn like humans.*
- *Types of AI:*
  - *Narrow/Weak AI: Designed for specific tasks (e.g., virtual assistants, game-playing AI)*
  - *General/Strong AI: Hypothetical AI with human-like cognitive abilities*
  - *Superintelligent AI: Theoretical AI surpassing human intelligence across all domains*

2. Historical Overview
- *1950s: Alan Turing proposes the Turing Test*
- *1956: Dartmouth Conference coins the term "Artificial Intelligence"*
- *1960s-70s: Early AI winters due to limitations in computing power*
- *1980s-90s: Expert systems and neural networks gain traction*
- *2000s-present: Big data, improved algorithms, and increased computing power lead to significant AI advancements*

3. Key Concepts and Technologies
- *Machine Learning: Algorithms that improve through experience*
  - *Supervised Learning*
  - *Unsupervised Learning*
  - *Reinforcement Learning*
- *Deep Learning: Subset of machine learning using artificial neural networks*

- *Natural Language Processing (NLP): Enabling machines to understand and generate human language*
- *Computer Vision: Enabling machines to interpret and understand visual information*
- *Robotics: Integration of AI in physical machines to perform tasks*

4. Current Applications
- *Virtual assistants (e.g., Siri, Alexa)*
- *Recommendation systems (e.g., Netflix, Amazon)*
- *Autonomous vehicles*
- *Healthcare diagnostics and drug discovery*
- *Financial trading and fraud detection*
- *Content creation (text, images, music)*
- *Industrial automation and predictive maintenance*

5. Ethical Considerations and Challenges
- *Bias and fairness in AI systems*
- *Privacy concerns and data protection*
- *Job displacement due to automation*
- *Accountability and transparency in AI decision-making*
- *Potential misuse of AI (e.g., deep fakes, autonomous weapons)*

6. Future Prospects
- *Continued advancements in natural language understanding and generation*
- *Improved human-AI collaboration*
- *Potential breakthroughs in general AI*
- *Integration of AI with other emerging technologies (e.g., IoT, blockchain)*
- *Expansion of AI applications in climate change mitigation, education, and scientific research*

7. Key Players and Organizations
- *Tech giants: Google, Microsoft, IBM, Amazon,*

Facebook
- *AI research institutions: OpenAI, DeepMind, MIT, Stanford AI Lab*
- *International bodies: IEEE, Partnership on AI, AI4ALL*

8. Learning Resources
- *Online courses: Coursera, edX, Udacity*
- *Books: "Artificial Intelligence: A Modern Approach" by Stuart Russell and Peter Norvig*
- *Conferences: NeurIPS, ICML, AAAI*
- *Journals: Journal of Artificial Intelligence Research, AI Magazine*

9. Key AI Terms
- *Algorithm: A set of rules or instructions given to an AI program to help it learn on its own*
- *Neural Network: A computer system modeled on the human brain and nervous system*
- *Machine Learning: An application of AI that provides systems the ability to automatically learn and improve from experience*
- *Deep Learning: A subset of machine learning based on artificial neural networks*
- *Natural Language Processing (NLP): The ability of a computer program to understand human language as it is spoken*
- *Robotics: The branch of AI that deals with the design, construction, operation, and use of robots*
- *Computer Vision: A field of AI that trains computers to interpret and understand the visual world*
- *Websites*
  1. www.commonsense.org
  2. www.childtrends.org
  3. https://innovationschoolchoice.com/artificial-intelligence-ai-parent-resources/
  4. https://www.healthychildren.org/English/family-life/Media/Pages/how-will-artificial-intelligence-

AI-affect-children.aspx

5. https://travelanewpath.com/parenting-in-the-age-of-ai

ABOUT THE AUTHOR

Lorne is the senior director of International Sales and Training for FIA Coaching as well as a lifelong entrepreneur. FIA has successfully trained over 11,000 coaches in 90+ countries. Lorne is the Sr. Director of International Sales and Training for FIA Coaching and is spearheading the company's AI coaching initiative.

His background ranges from insurance, franchising, career transition, Internet marketing, feature film financing, celebrity golf events, and coaching.

Lorne is also the founder of the AI Virtual Trade Show that will soon feature the top AI companies in the world in a one of a kind AI Virtual Showroom.

Lorne is married to his wife Elisa and lives in Murfreesboro, Tennessee. Together they have seven grown children and ten grandchildren. To contact Lorne please email directly at Lorne@fiacoaching.com.

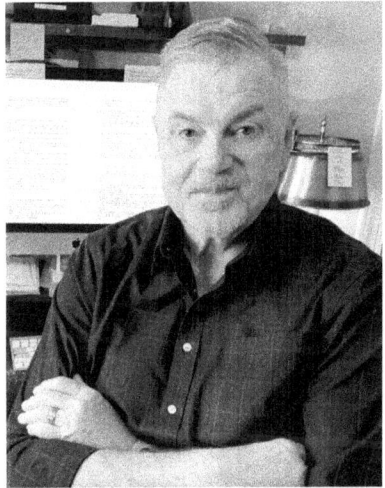

BONUS TO PURCHASERS
OF THIS BOOK

Because AI is changing so rapidly it will be necessary to write revisions to this book. To thank my readers, I am offering you free ebook updates on *"AI-Careful Thinking."* I am doing this to keep you all "careful," and we know in order to be"careful" you must be informed, So, any updates to this book will be sent to my readers for free in ebook format. In order for you to receive your free ebook I need your email address, so I know where to send the ebook updates (send your email address to Lorne@bookjolt.com.)

P.S. Could you do me a favor? Would you write a review of the book? It would be greatly appreciated!

GLOSSARY OF AI TERMS

A

Artificial Intelligence (AI): The simulation of human intelligence in machines that are programmed to think and learn like humans.

Algorithm: A set of rules or instructions given to an AI, neural network, or other machine to help it learn on its own.

Artificial Neural Network (ANN): A computing system inspired by biological neural networks that constitute animal brains.

B

Big Data: Extremely large datasets that may be analyzed computationally to reveal patterns, trends, and associations.

Backpropagation: An algorithm for supervised learning of artificial neural networks using gradient descent.

C

Computer Vision: A field of AI that trains computers to interpret and understand the visual world.

Chatbot: A computer program designed to simulate conversation with human users, especially over the Internet.

D

Deep Learning: A subset of machine learning based on artificial neural networks with multiple layers.

Data Mining: The process of discovering patterns in large datasets involving methods at the intersection of machine learning, statistics, and database systems.

E

Expert System: An AI system that emulates the decision-making ability of a human expert.

Evolutionary Computation: A family of algorithms for global optimization inspired by biological evolution.

F

Fuzzy Logic: A form of many-valued logic in which the truth values of variables may be any real number between 0 and 1.

Feature Extraction: The process of reducing the number of resources required to describe a large set of data accurately.

G

Generative AI: AI systems that can generate new content, such as text, images, or audio, based on training data.

Genetic Algorithm: A search heuristic that mimics the process of natural selection.

H

Heuristic: A technique designed for solving a problem more quickly when classic methods are too slow, or for finding an approximate solution when classic methods fail to find any exact solution.

I

Intelligent Agent: An autonomous entity which observes through sensors and acts upon an environment using actuators and directs its activity towards achieving goals.

K

Knowledge Base: A technology used to store complex structured and unstructured information used by a computer system.

M

Machine Learning: The study of computer algorithms that improve automatically through experience.

Model: In machine learning, a model is a specific representation learned from data.

N

Natural Language Processing (NLP): The ability of a computer program to understand human language as it is spoken.

Neural Network: A series of algorithms that endeavors to recognize underlying relationships in a set of data through a process that mimics the way the human brain operates.

O

Overfitting: When a statistical model fits exactly against its training data but fails to fit additional data or predict future observations reliably.

P

Perceptron: An algorithm for supervised learning of binary classifiers.

Pattern Recognition: The automated recognition of patterns and regularities in data.

R

Reinforcement Learning: An area of machine learning concerned with how software agents ought to take actions in an environment so as to maximize some notion of cumulative reward.

Robotics: The branch of technology that deals with the design, construction, operation, and application of robots.

S

Supervised Learning: The machine learning task of learning a function that maps an input to an output based on example input-output pairs.

Sentiment Analysis: The use of natural language processing to systematically identify, extract, quantify, and study affective states and subjective information.

T

Transfer Learning: A research problem in machine learning that focuses on storing knowledge gained while solving one problem and applying it to a different but related problem.

Turing Test: A test of a machine's ability to exhibit intelligent behavior equivalent to, or indistinguishable from, that of a human.

U

Unsupervised Learning: A type of machine learning that looks for previously undetected patterns in a dataset with no pre-existing labels and with a minimum of human supervision.

V

Virtual Assistant: An application program that understands natural language voice commands and completes tasks for the user.

www.ingramcontent.com/pod-product-compliance
Lightning Source LLC
Chambersburg PA
CBHW060607200326
41521CB00007B/691